Clinical and Genetic Aspects of Sudden Cardiac Death in the Practice of Sports Medicine

Colloquium Series on
Genomic and Molecular Medicine

Editor
Dhavendra Kumar
The University of Glamorgan
UK Institute of Medical Genetics
Cardiff University School of Medicine
University Hospital of Wales

The progress of medicine has always been driven by advances in science and technology. The practice of medicine at a given place and time is a reflection of the current knowledge, applications of the available information and evidence, social/cultural/ religious beliefs, and statutory requirements. Thus, it needs to be dynamic and flexible to accommodate changes and new developments in basic and applied science in keeping with the individual and societal expectations. From 1970 onwards, there has been a continuous and growing recognition of the molecular basis of medical practice. Most medical curricula allow sufficient space and time to ensure satisfactory coverage of the basic principles of molecular biology. Emphasis is given on the relevance of molecular science in the practice of clinical medicine paving the way for a more holistic approach to patient care utilizing new dimensions in diagnostics and therapeutics. It is extremely important that both teachers and students have an agreed agenda for learning and applications of molecular medicine. This should not be restricted to few uncommon genetic conditions but extended to include inflammatory conditions, infectious diseases, cancer, and age-related degenerative conditions involving multiple body systems.

Alongside the developments and progress in molecular medicine, rapid and new discoveries in genetics led to an entirely new approach to the practice of clinical medicine. However, the field of genetic medicine has been restricted to the diagnosis, offering explanation and assistance to patients and clinicians in dealing with a number of relatively uncommon inherited disorders. Nevertheless, this field has gradually established and accorded the specialty status in the medical curriculum of several countries.

Clinical and Genetic Aspects of Sudden Cardiac Death in the Practice of Sports Medicine
Lynne Millar, Nabeel Sheikh, and Sanjay Sharma
www.morganclaypool.com

ISBN: 9781615043866   paperback

ISBN: 9781615043873   ebook

DOI: 10.4199/C00062ED1V01Y201206GMM002

A Publication in the

COLLOQUIUM SERIES ON GENOMIC AND MOLECULAR MEDICINE

Lecture #2

Series Editors: Dhavendra Kumar, Cardiff University School of Medicine, University Hospital of Wales

Series ISSN

ISSN   2167-7840   print

ISSN   2167-7859   electronic

Clinical and Genetic Aspects of Sudden Cardiac Death in the Practice of Sports Medicine

Lynne Millar
Nabeel Sheikh
Sanjay Sharma
St. George's University of London,
United Kingdom

COLLOQUIUM SERIES ON GENOMIC AND MOLECULAR MEDICINE #2

ABSTRACT

Sudden cardiac death is the leading cause of non-traumatic mortality in young (<35 years old) athletes, with recent data suggesting the incidence to be higher than what was previously estimated. The vast majority of deaths are caused by silent hereditary or congenital cardiac disorders. Over the last decade, advances in our understanding of both the genetic and clinical mechanisms underlying these conditions, particularly those associated with a structurally normal heart, have led to advances in diagnosis and management including interventions and lifestyle modifications that aim to minimize the risk of sudden cardiac death (SDC). Coupled with effective screening programs, other strategies such as emergency response planning and the use of automated external defibrillators have also emerged as strategies in preventing and treating sudden cardiac arrest.

This book aims to provide an overview of the genetic and clinical aspects of SCD in young athletes, with particular emphasis on the specific issues related to diagnosis and management that these unique group of individuals pose to a physician. Specific diagnostic and management dilemmas will be illustrated through clinical cases and the most up-to-date guidelines regarding participation in sport outlined.

KEYWORDS

sudden cardiac death, athlete's heart, genetics of inherited cardiac conditions, cardiomyopathies, hypertrophic cardiomyopathy, arrhythmogenic right ventricular cardiomyopathy, ion channel disorders, long-QT syndrome, brugada syndrome, congenital coronary arteries, pre-participation screening

Contents

Contents

List of Abbreviations

ACE	Angiotensin-converting enzyme
AED	Automated external defibrillator
AF	Atrial fibrillation
ARVC	Arrhythmogenic right ventricular cardiomyopathy
AV	Atrioventricular
BAV	Bicuspid aortic valve
BrS	Brugada syndrome
Ca^{2+}	Calcium
CCAA	Congenital coronary artery anomalies
CMRI	Cardiac Magnetic Resonance Imaging
CO	Cardiac output
CPVT	Catecholaminergic polymorphic ventricular tachycardia
CPEX	Cardiopulmonary exercise test
DAD	Delayed after-depolarization
DCM	Dilated cardiomyopathy
ECG	Electrocardiogram
ER	Early repolarization
HCM	Hypertrophic cardiomyopathy
ICD	Implantable cardioverter–defibrillator
K^+	Potassium
LMCA	Left main coronary artery
LQTS	Long-QT syndrome
LV	Left ventricle
LVH	Left ventricular hypertrophy
LVOTO	Left ventricular outflow tract obstruction
LVWT	Left ventricular wall thickness
MSEC	Milliseconds
MVP	Mitral valve prolapse

Na^+	Sodium
NSVT	Non-sustained ventricular tachycardia
NYHA	New York Heart Association
PCCD	Progressive cardiac conduction defect
PPSP	Pre-participation screening program
QTc	Corrected QT interval
RCA	Right coronary artery
RV	Right ventricle
RVOT	Right ventricular outflow tract
SADS	Sudden arrhythmic death syndrome
SCA	Sudden cardiac arrest
SCD	Sudden cardiac death
SV	Stroke volume
TGF-β	Transforming growth factor beta
TWI	T-wave inversion
VF	Ventricular fibrillation
VT	Ventricular tachycardia
WPW	Wolff–Parkinson–White

CHAPTER 1

Introduction

The health benefits of regular physical exercise are well established. Apart from improving stamina and general well-being, participation in regular exercise aids weight reduction, improves lipid and blood pressure profile, and increases insulin sensitivity that collectively reduces the risks of coronary atherosclerosis. Additionally, exercise may play a protective role against the development of certain cancers such as prostate and colon [1–3]. Indeed, individuals who exercise regularly live an average of 6–7 years more than sedentary persons. Based on the extent of physical activity they perform, athletes are therefore regarded by society as the epitome of health. Yet paradoxically, in a small proportion, vigorous exercise may trigger sudden cardiac death (SCD) [4–12]. Although rare, such events are catastrophic not only because of the number of life years lost in a usually young, asymptomatic, healthy individual but also because they frequently occur in the public domain. The resulting media attention afforded to such tragedies, the ensuing public scrutiny, and concerns raised within the medical and sporting communities invariably trigger debate centered on the prevention of such deaths in the future [13–16].

Sudden cardiac death is the leading cause of non-traumatic mortality in young (aged <35 years) athletes [9, 12, 17–19]. It is defined as an unexpected, natural death secondary to a cardiac cause within 1 hour from the onset of symptoms in an individual without a previously diagnosed cardiac abnormality [20]. Most cases are due to a silent, hereditary or congenital cardiac condition that is typically the trigger for a fatal ventricular arrhythmia [7, 12, 19–21]. In older athletes (>35 years), the majority of cases are due to atherosclerotic coronary artery disease [22–24].

In recent years, there have been numerous advances in our understanding of both the genetic and clinical aspects of disorders implicated in SCD in athletes. Such knowledge has led to the development of numerous therapeutic interventions and lifestyle modifications, which aim to minimize the risk of SCD in this unique group of individuals. Based on these considerations and increased awareness of SCD in young athletes, sporting bodies have implemented pre-participation screening programs (PPSPs) in an attempt to identify those athletes potentially at risk [5, 25–27]. In addition, given that no screening program can offer complete protection, other strategies including education regarding correct recognition and prompt treatment of sudden cardiac arrest (SCA) have also emerged [28–30].

The aim of this article is to provide an overview of the molecular and clinical aspects of SCD in athletes, focusing predominantly on young individuals (<35 years old). A brief description of incidence and demographics will be followed by an outline of normal cardiac adaptation to exercise (the "athlete's heart"), knowledge of which is particularly important for the clinical assessment of an athlete. Conditions responsible for SCD will be discussed in detail, followed by a practical guide to the assessment and management of an athlete and an illustration of clinical dilemmas with case studies. Finally, the role of PPSPs and automated external defibrillators (AEDs) in the prevention of SCD will be outlined.

CHAPTER 2

Demographics and Incidence

The vast majority of SCDs in athletes occur either during or immediately after intense physical activity [10], indicating that exercise is a trigger for fatal arrhythmias in predisposed individuals. The metabolic stresses associated with exercise including surges in catecholamines, dehydration, increased myocardial oxygen demand, changes in blood pH, and electrolyte imbalance potentially explain the link between exercise and SCD. Certain cohorts also appear to be more susceptible; several studies indicate males to be at higher risk than females, with ratios varying from 2.6:1 to 9:1 [12, 17, 18, 28, 31, 32], although this may be due to higher participation rates in certain sports (such as soccer and American football) [19, 33] or reporting bias by the media. Some data indicate that in those aged <35 years, adolescent and younger athletes (i.e., <18 years old) are at highest risk [5, 12]. However, in cohorts aged >35 years, risk increases as coronary artery disease becomes an important factor [9, 23].

Data from the US indicate that race has an important bearing on an individual's risk of SCD, with athletes of African/Afro-Caribbean origin (black athletes) being at higher risk than Caucasians (white athletes) [34], particularly from hypertrophic cardiomyopathy (HCM) [17, 35]. Finally, an athlete's sporting discipline may also have an influence, with stop–start sports such as basketball, American football, and soccer being most commonly associated with SCD [12, 17] (although again, this may simply reflect higher participation rates in these sports). A recent study of the National Collegiate Athletic Association (NCAA) student athletes in the US demonstrated an SCD rate of 1 in 11,394 for basketball with death rates as high as 1 in 6993 in males and 1 in 5743 in African Americans [17]. This compares with an SCD incidence of 1 in 38,497 for American football and 1 in 43,770 for the study as a whole.

Several attempts have been made to define the precise incidence of SCD in athletes, but the precise figure varies depending on the cohort studied. Estimates from different studies vary widely (Table 1) [9, 12, 17–19, 28, 36–40] due to differences in the definition of SCD, the demographics of study populations (with certain cohorts being more at risk), and inaccuracies in determining the precise number of SCDs (numerator) and the total athletic population at risk (denominator).

In terms of defining the numerator and denominator, the most accurate method would be through carefully complied systematic registries of both the number of athletes experiencing SCD

TABLE 1: Widely varying estimates of the incidence of SCD in young (aged <35 years) athletes and children (see individual studies), comparing populations and methodologies used to derive numerators and denominators. Taken from Sheikh and Sharma [41], with permission.

| STUDY | POPULATION | AGE RANGE, y | METHODS AND REPORTING SYSTEM FOR NUMERATOR AND DENOMINATOR | | INCIDENCE |
			NUMERATOR	DENOMINATOR	
Van Camp et al. [5]	High school and college athletes (US)	13–24	Public media reports and other reported cases/ newspaper clipping service	Estimated from participation rates in high school and college athletic associations	1:300 000
Maron et al. [18]	High school athletes (Minnesota, US)	13–19	Catastrophic insurance claims	Mandatory indemnity registry	1:200 000
Eckart et al. [51]	Military recruits (US)	18–35	Mandatory, autopsy-based	Military recruit enlistment data	1:9000
Drezner et al. [102]	College athletes (US)	18–23	Retrospective survey of head athletic trainers and use of AEDs	Intercollegiate athletes per institution reported by the athletic trainer	1:67 000
Corrado et al. [14]	Competitive athletes (Italy)	12–35	Mandatory registry for SCD	Mandatory participation registry	1:25 000
Maron et al. [17]	Competitive athletes (US)	12–35	Public media reports and other electronic databases	Estimated form participation rates in high school and college	1:166 000

Study	Population	Age range	Study design	Data source	Incidence
Atkins et al. [13]	Adolescents and young adults (US and Canada)	12–24	Prospective, population-based EMS reports	Estimated from US (2000) and Canadian (2001) census data	1:27 000
Chugh et al. [111]	Children (Oregon, US)	10–14	Prospective, population-based EMS/hospital reports	Calculated from total county population using US census year (2000) data	1:58 000
Drezner et al. [15]	High school athletes (US)	14–17	Cross-sectional survey	National registry for AED use containing data on number of athletes per high school	1:23 000
Solberg et al. [19]	Competitive athletes and physically active adults (Norway)	15–34	Retrospective review of mandatory national forensic registry	Estimated from the population of men in Norway in middle of study and a national health survey on physical activity	1:111 000
Harmon et al. [1]	NCAA college athletes (US)	17–24	Resolutions database, public media reports, and catastrophic insurance claims	Published NCAA records of the number of athletes participating each year	1:44 000

Abbreviations: AED, automated external defibrillator; EMS emergency medical service; NCAA Collegiate Athletic Association; SCD, sudden cardiac death.

per year and the total athletic population at risk. Unfortunately, such systems of recording deaths and participation rates are not available in most countries. The scientific literature has therefore relied on imprecise methods for collecting these data such as media reports, insurance claims, and cross-sectional surveys in the case of the number of SCDs, and average participation rates, cross-sectional surveys, and census data for estimation of the total population at risk (Table 1). A recent study highlighted the limitations of these methods, which have the potential to both over- or underestimate the true incidence: searching a combination of 2 databases systematically recording SCDs identified 87% of cases over a 5-year period compared to only 56% when rigorous electronic interrogation of media reports was employed [17].

Despite these limitations, several recent studies based on sound methodology have estimated the incidence of SCD in young athletes to be between 1 in 20,000 and 1 in 45,000, dispelling the notion that these are incredibly rare events [17, 28, 36, 37].

●　●　●　●

CHAPTER 3

Physiology of the Athlete's Heart

It is well established that regular participation in intensive physical activity results in several electrical, structural, and functional cardiac adaptations that together constitute the "athlete's heart." These physiological changes are fundamental for generating and maintaining the increase in cardiac output (CO) required to meet the demands of repeated bouts of intense physical activity. However, occasionally, the manifestation of athlete's heart may overlap with that observed in patients with morphologically mild expressions of several inherited cardiac conditions, underscoring the importance of correct interpretation.

3.1   CARDIOVASCULAR ADAPTATION TO EXERCISE

Cardiac output, the product of heart rate and stroke volume (SV), can rise by as much as 5- to 6-fold during intense exercise. Initially, heart rate is the main determinant of this rise and increases as a result of sympathetic activation and sustained parasympathetic withdrawal. However, maximal heart rate is intrinsic to an individual, decreases with age, [42] and does not increase with exercise training [43]. This is in contrast to stroke volume, which increases with regular intensive physical activity as a result of a combination of enhanced LV filling, increase in LV end-diastolic volume, and reduction in LV end-systolic volume. It is this increase in stroke volume that is the predominant mechanism by which athletes are able to sustain an increased cardiac output for prolonged periods of time. [44]

3.2   DETERMINANTS OF CARDIOVASCULAR ADAPTATION TO EXERCISE

Cardiovascular adaptation to exercise is determined by several factors, as illustrated in Figure 1 below. One key factor is the sporting discipline of the athlete and to which degree isotonic exercise (endurance training) is performed as compared to isometric exercise (strength training). In reality, an overlap usually exists between the 2 in many sporting disciplines. Endurance training (such as long-distance running, swimming, rowing, and cycling) predominantly creates a volume load on the LV and therefore chamber enlargement. In contrast, strength training (which is required for sports

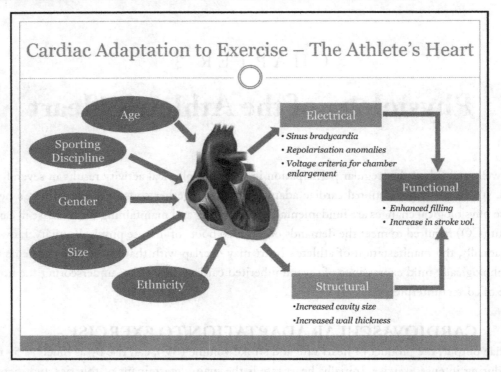

FIGURE 1: Diagram illustrating factors influencing cardiac adaptation to exercise, and phenotypic manifestations of the athlete's heart.

such as weightlifting, American football, and track and field throwing events) results in an increase in peripheral vascular resistance, transient hypertension, and a pressure load on the LV leading to wall hypertrophy.

An athlete's size and gender also influence cardiovascular adaptation to exercise, with large individuals and those of male gender developing the greatest increases in cardiac dimensions [45–47]. Adult athletes develop greater changes than adolescents of similar age and sporting discipline. [48, 49] Ethnicity has emerged as an important factor with respect to the manifestations of the athlete's heart. Several studies have demonstrated a higher prevalence of repolarization anomalies and greater magnitude of LV hypertrophy in black athletes compared with white athletes of similar age and sporting discipline [50–53]. This important aspect will be discussed in greater detail in section 3.3 below.

In recent years, several studies have indicated that genetic factors may have an important role in determining cardiovascular adaptation to exercise [54–62]. In particular, associations have been found between the presence (insertion allele, I) or absence (deletion allele, D) of a 287-base pair marker in the angiotensin-converting enzyme (ACE) gene and the extent of LV remodelling in response to physical training. Montgomery et al. [54] and subsequent investigators [55–61] found a significantly greater increase in LV mass in response to intensive exercise in individuals with the DD or DI allele. Changes in LV mass in response to athletic training have also been associated with the presence of polymorphisms in other common allelic variants of the renin–angiotensin system (RAS), such as angiotensinogen M235T and angiotensin type 1 recpetor, [62] with individuals possessing the TT and CC genotypes (respectively) showing the greatest increase in LV mass. These findings have been shown to occur regardless of gender, sporting discipline, and age; for example, a study in adolescent athletes demonstrated a greater (+28%) LV mass in individuals with the DD or ID alleles as opposed to the II allele [57].

The potential mechanisms by which the RAS may modulate LV remodelling in response to athletic training can be envisaged if one considers findings from studies demonstrating that angiotensin II promotes hypertrophy of cardiac myocytes through binding to angiotensin type 1 (AT1) receptors, [63, 64] which are present in heart tissue [65–67]. Angiotensin II also degrades kinins, which are known to inhibit myocyte growth. [68] Given that exercise activates ACE, this results in increased circulating levels of angiotensin II. Both the D allele of the ACE gene and T allele of the angiotensinogen gene are also associated with higher levels of angiotensin [61].

Besides the RAS, recent studies have implicated other genetic factors in cardiac remodelling and the development of LVH in humans. Using immunohistochemistry and Northern blotting techniques, the expression of insulin-like growth factor 1 (IGF-1) has been found to be increased in both animal models of cardiac hypertrophy and humans with LVH, [69–72] potentially implicating the cardiac IGF-1 gene in this process [73]. The mechanism by which this occurs is uncertain, but one possibility is through the effects that IGF-1 has on cell signalling via the phosphatidylinsitol 3 kinase–Akt1 pathway, [74] which is involved in the regulation of transcription factors and gene product synthesis [75].

3.3   THE ATHLETE'S HEART

Imaging studies involving both echocardiography and cardiac magnetic resonance imaging (CMRI) have demonstrated that approximately 50% of athletes develop structural cardiac changes in response to athletic conditioning, which include ventricular hypertrophy and chamber enlargement [46, 49, 76–81]. Electrocardiographic changes are also common and observed in up to 80% of athletes [82]. In rare instances, both the electrocardiographic and echocardiographic changes may overlap with the phenotypic expression of morphologically mild cardiac conditions implicated in SCD in young athletes. Thus, the differentiation between physiological changes and cardiac pathology is crucial since an erroneous diagnosis has the potential for serious consequences: an erroneous diagnosis of athlete's heart in an individual with cardiomyopathy may jeopardize a young life, and conversely, an erroneous diagnosis of cardiomyopathy in an athlete exhibiting extreme physiological manifestations of athlete's heart may lead to unfair disqualification from competitive sport.

3.3.1   Electrocardiographic Features

It was in 1918 that the famous American, Boston-based cardiologist Paul Dudley White published a paper entitled "The Pulse after a Marathon Race," noting a marked bradycardia in 4 endurance athletes and suggesting that this may be a normal, physiological finding in long-distance runners [83]. Since then, several electrocardiographic changes related to athletic conditioning have been established (Table 2; Figure 1), the majority of which are secondary to physiological adaptation resulting from increased resting parasympathetic tone, decreased sympathetic tone, and cardiac remodelling [84, 85]. Data derived from observational studies based on large athletic cardiovascular screening programs involving >32,000 unselected athletes [86] have enabled the identification of common physiological ECG changes that occur with athletic training from those that may be indicative of

TABLE 2: Common and uncommon ECG findings in athletes. Abbreviations: ECG— electrocardiogram; LBBB—left bundle branch block; RBBB—right bundle branch block. Adapted from Corrado et al. [82] and Uberoi et al. [97]	
COMMON AND TRAINING-RELATED ECG CHANGES	**UNCOMMON AND TRAINING-UNRELATED ECG CHANGES**
Sinus bradycardia	T-wave inversion > I mm
First-degree atrioventricular block	• in leads other than III, a VR, and V_1 and V_2 in white athletes
Incomplete RBBB Early repolarization Isolated QRS voltage criteria for left ventricular hypertrophy	• in leads other than III, aVR, and V_{1-4} (particularly if preceded by convex ST-segment elevation) in black athletes ST-segment depression Pathological Q waves Left atrial enlargement Left-axis deviation/left anterior hemiblock Right-axis deviation/left posterior hemiblock Right ventricular hypertrophy Ventricular preexcitation Complete LBBB or RBBB Long- or short-QT interval Brugada-like early repolarization

Abbreviations: ECG, electrocardiogram; LBBB, left bundle branch block; RBBB, right bundle branch block.

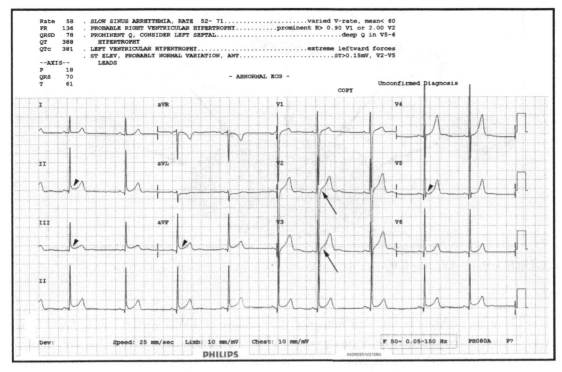

```
Rate    58   . SLOW SINUS ARRHYTHMIA, RATE  52~ 71.....................varied V-rate, mean< 60
PR     136   . PROBABLE RIGHT VENTRICULAR HYPERTROPHY..........prominent R> 0.90 V1 or 2.00 V2
QRSD    78   . PROMINENT Q, CONSIDER LEFT SEPTAL..............................deep Q in V5-6
QT     388       HYPERTROPHY
QTc    381   . LEFT VENTRICULAR HYPERTROPHY...........................extreme leftward forces
             . ST ELEV, PROBABLY NORMAL VARIATION, ANT.......................ST>0.15mV, V2-V5
  --AXIS--       LEADS
P       18
QRS     70                          - ABNORMAL ECG -
T       61
                                                   COPY              Unconfirmed Diagnosis
```

FIGURE 2: ECG from a professional athlete, demonstrating common, training related changes including sinus bradycardia, Sokolow–Lyon voltage criteria for LVH, ST-segment elevation (arrows) and early repolarization (arrowheads).

cardiac pathology. Common alterations include sinus bradycardia and first-degree atrioventricular (AV) block, with a small number of athletes exhibiting Mobitz type 1 second-degree AV block or a nodal rhythm at rest, which revert to sinus rhythm with exertion [87]. Isolated voltage criteria for LVH are present in up to 80% of athletes, and in the vast majority of cases, these may simply reflect physiological left ventricular enlargement [84, 88–92]. Isolated ECG criteria for LVH correlate poorly with LV mass and are observed in less than 2% of patients with HCM [50, 93, 94]. Repolarization changes, characterized by ST-segment and J point elevation (Figure 2) with high amplitude T-waves are found in up to 50%–80% of athletes [95, 96]. Such repolarization anomalies are particularly common in endurance athletes and black athletic individuals. In contrast, several uncommon ECG changes are occasionally observed in less than 5% of athletes, which are unrelated to training and warrant further investigations (Table 2). These include pathological Q waves, resting ST-segment depression, complete bundle branch block (particularly left bundle branch block [LBBB]) and T-wave inversions (TWIs) [82, 97].

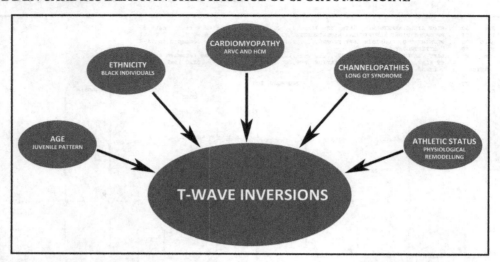

FIGURE 3: Several factors influence the presence of T-wave inversions on a resting ECG.

There are multiple factors that influence the presence of TWIs on the resting ECG (Figure 3). T-wave inversions may be normal manifestations in some athletic cohorts but are common findings in patients with both HCM [53] (Figure 4) and ARVC [98, 99]. Age and ethnicity have an important bearing on the presence of TWIs. They are a rare manifestation of athlete's heart in Caucasian individuals. T-wave inversions in leads V1–V2 are present in approximately 2% of all adolescent Caucasian athletes. Beyond V2, TWIs are observed in only 0.7% of Caucasian athletes under 16 and in 0.1% over 16 years old [100]. In contrast, TWIs may be present in up to 23% of male and 14% of female black athletes and are predominantly confined to the anterior precordial leads V1–V4 [51–53, 101, 102]. When present, they are usually associated with preceding convex ST-segment elevation (Figure 5).

Although rare (1%–2%), the significance of TWIs in the inferior and lateral leads is not fully recognized in white athletes; however, reports based on small groups harbouring such repolarization changes have revealed an association with an underlying cardiomyopathy or aborted SCD [103]. In contrast, TWIs in the inferior and lateral leads are present in a significant proportion of black athletes (6% and 4%, respectively). One report has made an association between this particular pattern and the diagnosis of HCM based on the finding in just 3 athletes [53]. Longitudinal studies, possibly involving assessment of 1st degree relatives of black athletes with inferior and/or lateral TWIs, are required to assess the precise significance of this repolarization pattern.

Based on these observations, current guidelines suggest that, in black athletes, TWIs preceded by ST-segment elevation in V1–V4 and, in white athletes, in V1–V2, in asymptomatic individuals without a significant family history do not require further investigation [82, 97]. However,

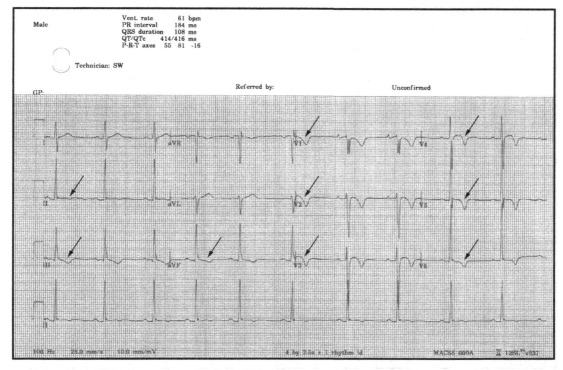

FIGURE 4: An ECG from a professional football player. Note the deep TWIs (arrows) beyond V4, extending into the lateral leads (V5, V6) and inferior leads (II, III, and aVF). Subsequent evaluation revealed phenotypic features of HCM.

TWIs in the inferior or lateral leads should always trigger comprehensive evaluation, particularly for HCM (Figure 5) irrespective of age or ethnicity [82, 97].

3.3.2 Echocardiographic Features

It has been recognized for over a century that athletic training is associated with increased cardiac dimensions. Indeed, it was the Swedish physician Henschen who coined the term "athlete's heart" back in 1899, after concluding that cross-country skiers exhibit both cardiac dilatation and hypertrophy and that both the right and left sides of the heart are enlarged [104]. Although Henschen used manual chest percussion as his diagnostic technique, it was not until the advent of chest radiography and then echocardiography that his conclusions were proven correct. Early studies using chest radiography confirmed global cardiac enlargement in trained athletes, exactly as postulated by Henschen [105–107]. With the subsequent advent of 2D echocardiography, preliminary M-mode studies in male athletes allowed quantification of these changes, demonstrating that, on average,

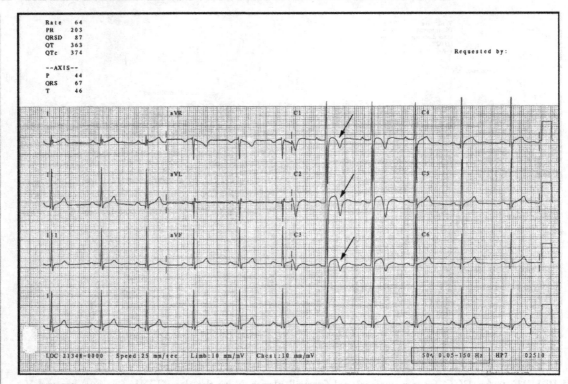

FIGURE 5: An ECG from a black athlete, showing deep TWIs in leads V1–V3 (arrows). Note preceding convex ST-segment elevation. Sokolow–Lyon voltage criteria for left ventricular hypertrophy and 1st degree AV-block are also present. Comprehensive evaluation and follow-up revealed no phenotypic features of a cardiomyopathy.

athletes developed a 15%–20% increase in LV wall thickness and 10% increase in end-diastolic diameter compared to the upper limits of normal in sedentary individuals [77, 108–111]. A study by Maron compared over 1000 male athletes and non-athletic controls, observing that both mean maximal left ventricular wall thickness (MLVWT) and end-diastolic diameters were significantly greater in the athletic group [77]. In 1976, Morganroth et al. [108] demonstrated not only that do athletes develop both increased wall thicknesses and chamber enlargement compared to controls but also that the extent of these is dependent on sporting discipline. Their observation that strength training resulted in predominantly concentric hypertrophy, whereas those undergoing endurance training demonstrated eccentric LV enlargement with cavity dilatation, leads to the concept of sport-specific cardiac remodelling, also known as the "Morganroth Hypothesis." In the modern era, echocardiographic and CMRI studies using advanced techniques have further quantified the extent of these changes and attempted to elucidate the functional adaptations that occur alongside structural cardiac remodelling [45, 46, 78–81]. Such studies have demonstrated that LV ejection fraction is generally normal among athletes [112] and that athletic training (and in particular endurance training) leads to enhanced early diastolic filling, [113–118] a factor that is crucial for maintaining stroke volume during intense physical activity [119].

Cardiac dimensions are determined by several factors including sex, size, age, sporting discipline, and ethnicity. Males in any given sport demonstrate larger cavities than females. For example, Pelliccia et al. [47] studied a cohort of 1309 predominately white, adult male and female athletes, observing larger LV end-diastolic diameters in males compared to females (Figure 6). Marked cavity dilatation of ≥60 mm was seen in 14% of their cohort, overlapping with the phenotype of dilated cardiomyopathy (DCM), although no individual exhibited LV cavity sizes of >70 mm. However, some athletes are capable of developing substantially large dimensions, particularly with respect to ventricular cavity size. Male athletes with large body surface area, particularly in endurance sports such as rowing, cycling, and running, exhibit the largest dimensions [120]. As expected, adults exhibit larger dimensions than adolescent athletes independent of age and ethnicity; observational studies in the adolescent population and black athletes by our group has allowed the determination of upper limits of normal LV cavity size in these populations (Table 3) [48, 49, 52].

A smaller proportion of athletes also develop substantial increases in LV wall thickness (LVWT). In another study by Pelliccia et al. [45], 947 white, adult male and female athletes were

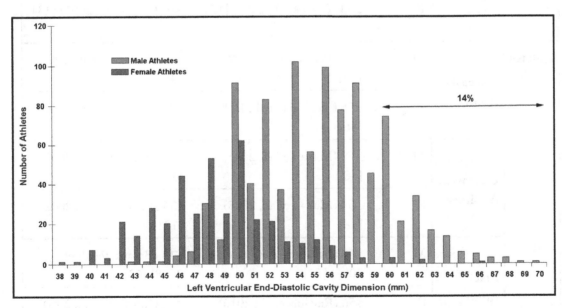

FIGURE 6: Distribution of left ventricular end-diastolic cavity dimensions in 1309 highly trained athletes without evidence of structural cardiovascular disease. Fourteen percent of athletes had markedly enlarged left ventricular cavities ranging in size from 60 mm to 70 mm. Adapted from Pelliccia et al. [46].

studied; wall thickness of ≥13 mm, which could be regarded in keeping with morphologically mild HCM, was rare (1.7%) and never exceeded 16 mm. Athletes with LVH almost always exhibited an enlarged LV cavity, and all showed normal diastolic function. All athletes with an LVWT ≥13 mm were males, with a mean age of 22.5 years, BSA of >2, and who participated in endurance sports. Similarly, a study in junior white athletes by Sharma et al. [121] found greater LV wall thicknesses in athletes compared to non-athletes; however, the magnitude was smaller than in adults, with only

TABLE 3: Upper reference limits* for left ventricular end-diastolic diameter, left ventricular wall thickness and right ventricular end-diastolic diameter in athletes and non-athletes. Adapted from Zaidi and Sharma, 2011 [134].

		GENDER	LEFT VENTRICULAR END-DIASTOLIC DIAMETER (MM)	LEFT VENTRICULAR WALL THICKNESS (MM)
Non-athletes		M	≤59 (Lang et al., 2005) [131]	≤10 (Lang et al., 2005) [131]
		F	≤53 (Lang et al., 2005) [131]	≤9 (Lang et al., 2005) [131]
Athletes	**Caucasian Adult**	M	≤63 (Pelliccia et al., 1999) [46]	≤12 (Pelliccia et al., 1991) [45]
		F	≤56 (Pelliccia et al., 1999) [46]	≤11 (Pelliccia et al., 1996) [47]
	Caucasian Adolescent	M	≤58 (Makan et al., 2005) [48]	≤12 (Sharma et al., 2002) [121]
		F	≤54 (Makan et al., 2005) [48]	≤11 (Sharma et al., 2002) [121]
	Black Adult	M	≤62 (Basavarajaiah et al., 2008) [52]	≤15 (Basavarajaiah et al., 2008) [52]
		F	≤56 (Rawlins et al., 2010) [50]	≤12 (Rawlins et al., 2010) [50]

* Upper reference limits are 95th percentile values (mean + 2 standard deviations), derived from the study referenced in each case.

a small proportion (0.4%) exhibiting an LV wall thickness exceeding upper limits of normal, which when present was accompanied with chamber enlargement.

All of the aforementioned studies were conducted in Caucasian athletes [45, 51, 122]. In the past few years, several studies in male and female black athletes have demonstrated a greater prevalence and magnitude of LVH compared to Caucasians, [49, 52, 53] with 18% of male black athletes developing a LV wall thickness >13 mm (compared to 4% of white athletes; Figure 7) [52] and 3% of female black athletes developing an LV wall thickness of >11 mm (compared to none of the white female athletes) [49]. Crucially, data from these studies have demonstrated that a small but significant proportion of male black athletes (up to 3%) may exhibit wall thicknesses of ≥15 mm

BASAL RIGHT VENTRICULAR END-DIASTOLIC DIAMETER (MM)	MID RIGHT VENTRICULAR END-DIASTOLIC DIAMETER (MM)	BASE-TO-APEX RIGHT VENTRICULAR END-DIASTOLIC DIAMETER (MM)
≤42 (Rudski et al., 2010) [132]	≤35(Rudski et al., 2010) [132]	≤86 (Rudski et al., 2010) [132]
≤42 (Rudski et al., 2012) [132]	≤35 (Rudski et al., 2010) [132]	≤86 (Rudski et al., 2010) [132]
≤45 (D' Andrea et al., 2011) [124]	≤40 (D' Andrea et al., 2011) [124]	≤87 (D' Andrea et al., 2011) [124]
≤42 (D' Andrea et al., 2011) [124]	≤38 (D' Andrea et al., 2011) [124]	≤86 (D'Andrea et al., 2011) [124]
Not known	Not known	Not known
Not known	Not known	Not known
Not known	Not known	Not known
Not known	Not known	Not known

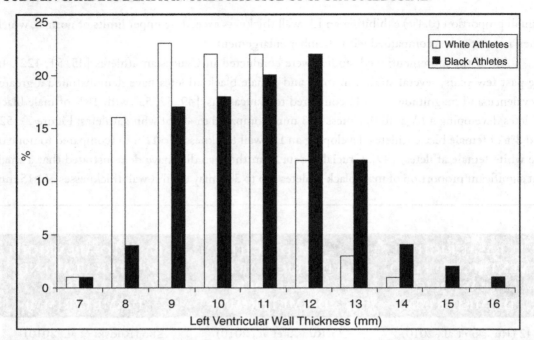

FIGURE 7: Distribution of maximal left ventricular wall thicknesses (MLVWT) in black and white athletes. Note the greater magnitude of LVH (MLVWT > 12 mm) in black athletes, including substantial LVH (MWVL ≥ 15 mm) in 3%. Taken from Basavarajaiah et al. [52].

[52]. This high prevalence of LVH in black athletes has important implications with respect to the differentiation between physiologic adaptation and HCM.

With data from black athletes now available, upper limits of normality for cardiac dimensions have been established in both black and white cohorts, summarized in Table 3.

Compared to the LV, remodelling in the right ventricle (RV) in response to athletic training has been less extensively studied. The crescent shape of the RV and increased myocardial trabeculations creates difficulty in determining which portion of the chamber to measure during echocardiography. Comparison with CMRI indicates that the best correlation between the 2 imaging modalities for RV end diastolic volumes exists with the proximal RVOT measured on m-mode in the parasternal long axis view (RVOT-Prox, Figure 8A), the RV long-axis diameter in the apical 4-chamber view (RVD3, Figure 8B) and the basal RV diameter in the apical 4-chamber view (RVD1, Figure 8B) [123].

Both the LV and RV are subject to equal preload during exercise; however, the RV may be exposed to a relatively higher afterload than the LV. In contrast to the systemic circulation, resistance in the pulmonary arterial vasculature decreases to a far lesser extent in response to exercise. As a consequence, because of the resultant increase in pulmonary artery pressure, the RV is subject to a far greater work load than the LV during physical activity.

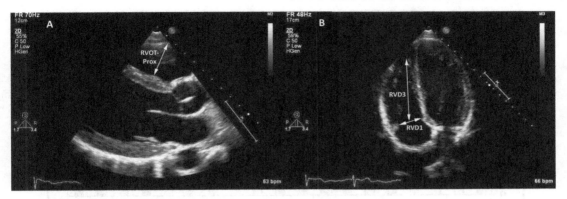

FIGURE 8: Areas of RV measurement on echocardiography that correspond most closely to CMRI RV end diastolic volumes.

Despite this, it appears that remodelling occurs in a "balanced" fashion, with similar changes observed in LV and RV mass, volume, and function in endurance athletes [80]. Echocardiographic limits of normality with respect to RV chamber size have recently been established for adult athletes (Table 3; Figure 9), although data are currently lacking for certain cohorts including black athletes and younger individuals.

Besides the ventricles, other cardiac structures may also undergo remodelling in response to intense athletic training. Numerous studies dating back to the 1980s have consistently demonstrated left atrial enlargement in highly trained athletes [125–129]. The largest study is from Pellicia et al. [127], who examined measurements in 1777 individuals, finding that 20% developed a diameter of >40 mm. Interestingly, only a small proportion of these individuals had clinical evidence of supraventricular arrhythmias.

Given the hemodynamic load placed upon it during exercise, it is not surprising that several studies have also reported aortic dimensions to be significantly larger in highly trained athletes compared to the general population [130, 131]. Debate, however, exists as to whether sporting discipline and the type of exercise performed have an influence. Although previous work has suggested that athletes performing strength training (which causes profound hypertension during exertion) develop significantly greater aortic dimensions than endurance-trained subjects [130, 131], a recent study by Pellicia et al. [132] revealed the opposite, with swimmers and cyclists exhibiting the largest increases. Whatever the role of sporting disciple is, an important agreement between these and other studies [133] is that the aortic root dimension rarely exceeds 40 mm through athletic training alone. Hence, any athlete with an aortic root dimension approaching or >40 mm requires careful assessment for the presence of any predisposing conditions (such as collagen disorders) in addition to long-term surveillance with serial echocardiography [132].

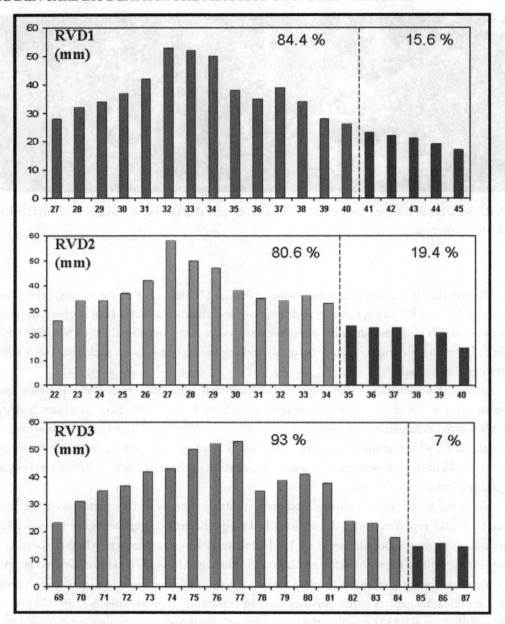

FIGURE 9: Distribution of right ventricular diameters in athletes. Dotted lines indicate the upper limits of diameters in controls. The percentage of athletes' RV diameters above and below such limits are also indicated. Adapted from D'andrea et al. [124].

CHAPTER 4

Etiology

In older athletes aged >35 years, most sport-related SCDs are due to atheromatous coronary artery disease [22–24]. In contrast, the vast majority of deaths in young athletes (<35 years old) are due to several inherited, congenital or acquired cardiac disorders, which in the vast majority (up to 80%) are usually silent up until the time of death [10, 31, 134]. As with differences in incidence, studies vary as to the commonest cause; some data from the United States (US) show HCM to be the commonest (Figure 10) [12, 19, 31, 135], accounting for approximately one third of cases, whereas other US data show congenital coronary artery anomalies (CCAA) to be the most frequent [18, 34]. In Italy, arrhythmogenic right ventricular cardiomyopathy (ARVC) is the most important cause [37], a similar finding to one study from Denmark [136]. In other European countries such as Norway, coronary artery disease appears to be important [9], whereas 1 autopsy-based study in the United Kingdom found cardiomyopathies to be the leading cause of death [137].

This section describes the cardiac conditions responsible for SCD in athletes in detail, focusing on genetic aspects followed by clinical considerations and case studies where appropriate. During the remainder of this article, we refer to the intensity of sports as classified by Mitchell et al. [138] and shown in Table 4.

4.1 THE CARDIOMYOPATHIES

The cardiomyopathies are some of the most important causes of SCD in young athletes, accounting for nearly half of all cases [31, 135]. Some US data suggest HCM to be the commonest cause, responsible for up to one third of all cases [12, 19, 31, 135], whereas data from Europe show ARVC to be commoner, accounting for one quarter of all deaths in some studies [37, 136].

4.1.1 Hypertrophic Cardiomyopathy

Hypertrophic cardiomyopathy is one of the most common inherited cardiac diseases occurring at a frequency of 1 in 500 in the general population, with a risk of sudden death of 1% per annum in those affected [139]. Previously being referred to as hypertrophic obstructive cardiomyopathy, the term "obstructive" has now been dropped reflecting the fact that 75% of patients do not have left ventricular outflow tract obstruction (LVOTO) at rest [140]. Microscopically, the condition is

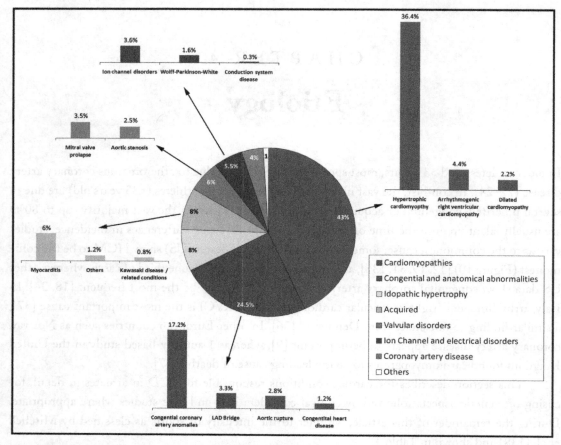

FIGURE 10: Causes of sudden cardiac death in young athletes. Data taken from US (Maron et al. [12]).

characterized by myocyte disarray (Figure 13D). Hypertrophy is due to down regulation of α-MHC and β-MHC induction, which represents a fetal state of MHC expression [141]. In 50%–60% of patients, it is inherited in an autosomal dominant fashion.

Hypertrophic cardiomyopathy is largely a disease of the sarcomere (Figure 11) [142], with the 2 most common mutations affecting the genes encoding myosin heavy chain 7 (MYH7) and myosin-binding protein C3 (MYPBC3), each making up 25% of cases (Table 5). Other common familial mutations include cardiac troponin T (TNNT2), cardiac troponin I (TNNI), α-tropomyosin (TPM1), α-actin (ACTC), myosin ventricular regulatory light chain (MYL2), and myosin ventricular essential light chain (MYL3) [139, 143–145]. In more recent years, mutations of α-myosin heavy chain (encoded by MYH6) [146] and cardiac troponin C (encoded by TNNC1)

TABLE 4: The Mitchell Classification of sports. This classification is based on peak static and dynamic components achieved during competition. It should be noted, however, that higher values may be reached during training. The increasing dynamic component is defined in terms of the estimated percent of maximal oxygen uptake (MaxO$_2$) achieved and results in an increasing cardiac output. The increasing static component is related to the estimated percent of maximal voluntary contraction (MVC) reached and results in an increasing blood pressure load. The lowest total cardiovascular demands (cardiac output and blood pressure) are shown in **green** and the highest in **red**. **Blue, yellow, and orange** depict low moderate, moderate, and high moderate total cardiovascular demands, respectively. *Danger of bodily collision. †Increased risk if syncope occurs. Taken from Mitchell et al. [138], with permission.

	A. Low (<40% Max O$_2$)	B. Moderate (40–70% Max O$_2$)	C. High (>70% Max O$_2$)
III. High (>50% MVC)	Bobsledding/Luge*†, Field events (throwing), Gymnastics*†, Martial arts*, Sailing, Sport climbing, Water skiing*†, Weight lifting*†, Windsurfing*†	Body building*†, Downhill skiing*†, Skateboarding*†, Snowboarding*†, Wrestling*	Boxing*, Canoeing/Kayaking, Cycling*†, Decathlon, Rowing, Speed-skating*†, Triathlon*†
II. Moderate (20–50% MVC)	Archery, Auto racing*†, Diving*†, Equestrian*†, Motorcycling*†	American football*, Field events (jumping), Figure skating*, Rodeoing*†, Rugby*, Running (sprint), Surfing*†, Synchronized swimming†	Basketball*, Ice hockey*, Cross-country skiing (skating technique), Lacrosse*, Running (middle distance), Swimming, Team handball
I. Low (<20% MVC)	Billiards, Bowling, Cricket, Curling, Golf, Riflery	Baseball/Softball*, Fencing, Table tennis, Volleyball	Badminton, Cross-country skiing (classic technique), Field hockey*, Orienteering, Race walking, Racquetball/Squash, Running (long distance), Soccer*, Tennis

Increasing Static Component ↑ — Increasing Dynamic Component →

[147] have been uncovered. Although rare, 2 mutations of the giant protein titin (encoded by TTN and which provides the molecular "spring" of cardiac muscle) have been discovered [148].

Genetic analysis has now expanded beyond the sarcomere-related genes, and new subgroups due to abnormalities of Z-discs and calcium-handling have been described. The first Z-disc mutations to be associated with HCM were in muscle LIM protein (responsible for the regulation of

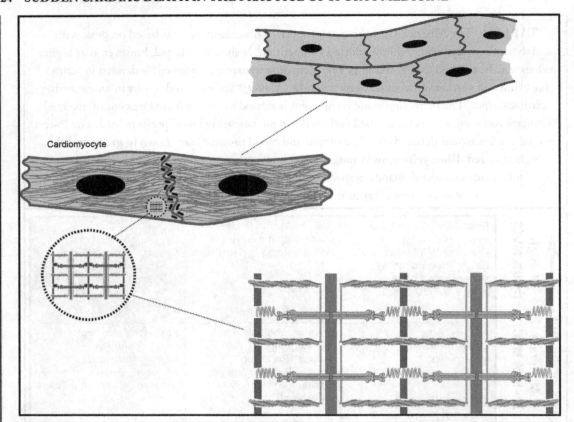

Cardiomyocyte

FIGURE 11: Schematic representation of the sarcomere and its relationship to the cardiomyocyte.

myogenic differentiation), encoded for by CSRP3 [149], and telethonin, encoded by TCAP [150]. Furthermore, mutations in LDB3-encoded LIM binding domain 3, ACTN2-encoded α-actinin 2, and VCL vinculin/metavinculin have also been discovered [151]. Nexilin, which is encoded by NEXN, is a cardiac Z-disc protein recently identified as a crucial component that functions to protect the Z-disc from forces generated by the sarcomere. Mutations in the NEXN gene are the most recent mutations to be described in HCM [152]. Although far less common, mutations in calcium handling causing HCM have been of interest. These include mutations in phospholamban, a protein that inhibits cardiomyocyte sarcoplasmic reticulum Ca^{2+}–ATPase (encoded for by PLN) [153] and mutations in junctophilin-2 (encoded for by the JPH2) (Table 5; Figure 12).

There are several metabolic disorders that may mimic HCM. Although they are rare, it is worth considering these conditions in patients with characteristics of HCM and features of multi-organ involvement. Fabry's disease, an X-linked condition due to a mutation in α-galactosidase A,

TABLE 5: Genetic mutations associated with HCM.

SACROMERIC GENE MURATIONS	
GENE	PROTEIN
MYH7	β-myosin heavy chain
MYH6	α-myosin heavy chain
MYL2	Regulatory light chain
MYL3	Essential light chain
MYBPC3	Myosin-binding protein C
TNNT2	Cardiac troponin T
TNNI3	Cardiac troponin I
TNNC1	Cardiac troponin C
TPM1	α-tropomyosin
ACTC	α-Actin
TTN	Titin
NON-SARCOMERIC PROTEINS	
Z-DISK PROTEINS	
LBD3	LIM binding domain 3(ZASP)
CSRP3	Muscle LIM protein
TCAP	Telethonin
VLC	Vinculin/metavinculin
ACTN2	α-Actinin 2

TABLE 5: (*continued*)	
CALCIUM RELEASE MUTATIONS	
RyR2	Calcium ryanodine receptor
JPH2	Junctophilin
PLN	Phospholambdan
METABOLIC HCM PHENOCOPIES	
PRKAG2	AMP-activated protein kinase
LAMP2	Lysosome-associated membrane
GLA	α-Galactosidase A
FXN	Frataxin

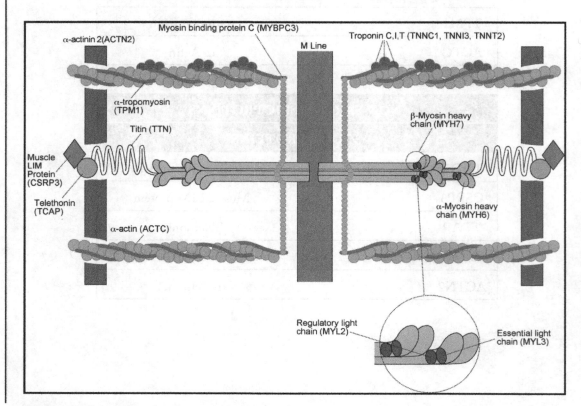

FIGURE 12: Schematic representation of sarcomeric and related mutations seen in HCM.

may express predominantly cardiac features and produce hypertrophy of the cardiac muscle, which is indistinguishable from HCM on transthoracic echocardiography and CMRI. Danon disease is a rare X-linked dominant condition due to a mutation in lysosome-associated membrane protein 2 (encoded by LAMP2). As well as showing cardiac hypertrophy akin to HCM, those affected have multisystem involvement with hepatic impairment, skeletal myopathy, and heart failure. PRKAG2-related cardiomyopathy is another glycogen storage disorder, which may mimic HCM. It is inherited in an autosomal dominant fashion and is due to an abnormality in the PRKAG gene, which normally encodes for AMP-activated protein kinase γ [154]. Importantly, both LAMP2- and PRAKG2-related cardiomyopathies can be distinguished from HCM by the presence of electrophysiological abnormalities including supraventricular arrhythmias, atrial fibrillation, and, in particular, ventricular pre-excitation on the resting ECG [155].

4.1.1.1 Genotype–Phenotype Correlations.   Although the yield of genotyping in HCM is only 60%–70%, identification of the causative gene may be useful in guiding management given that the severity of the phenotype and natural progression of the disease can be related to the type of mutation present [156–158]. For example, a study carried out by Olivotto et al. [159] looked at disease progression in patients with HCM with sarcomere mutations compared to non-sarcomere mutations, finding a more severe disease pattern in the former. Since this study has been published, more information has been discovered regarding mutations in Z-discs and calcium handling, which was not originally included; therefore, more research is needed to compare the natural progression of these branches of HCM. It has also been found that genotype tends to affect the age of onset of LVH. Those with MYH7 mutations often have pronounced LVH by the second decade of life, whereas patients who present with late-onset disease tend to have mutations in myosin-binding protein C, troponin I, and α-cardiac myosin heavy chain [160]. It should be noted that there are some genotype-positive patients who have normal LV thickness but who can still experience SCD. This is particularly true of those with troponin and myosin-binding protein C mutations [161]. Those with multiple mutations—so-called "compound heterozygotes"—appear to have a more severe phenotype. It is thought that double heterozygosity is present in approximately 5% of familial HCM [157, 162, 163]. Although rare, those with triple heterozygosity have an even more severe course, with a 14-fold increase in the risk of end stage disease and a high risk of ventricular arrhythmias [164].

4.1.1.2 Clinical Features and Management.   Hypertrophic cardiomyopathy is an extremely heterogeneous condition with a wide spectrum of clinical presentations. A murmur may be identified incidentally (or occasionally in an athlete at a pre-participation screening), or the patient may be picked up through screening on the basis of a positive family history. Alternatively, individuals may present with various symptoms to a physician or have an ECG for some unrelated reason that is noted to be abnormal. Rarely, SCD may be the first manifestation of the condition. Some of the earlier observational studies examining HCM portrayed it as a fairly malignant disorder, with an estimated incidence of SCD of up to 6%. Recent data would suggest that the incidence of SCD in these studies was over-estimated—it is likely that these studies were exposed to selection bias, being conducted predominantly in tertiary/specialist HCM centers and therefore including patients with a more severe phenotype [165].

The symptomatic presentation in patients is determined by the magnitude of LVH, diastolic dysfunction, myocardial ischemia, dynamic LVOTO, and cardiac arrhythmias. Some patients may have the condition without it being troublesome, and they may die from an unrelated condition. Chest pain can be a predominant feature in some patients, the etiology of which is 2-fold. First, there is thickening of the intimal and medial walls of the small vessels in the myocardium [166], leading to small vessel dysfunction. Second, this is compounded by the mismatch between myocardial mass and coronary circulation. As this myocardial ischemia is due to small vessel disease, coronary angiography is usually normal in these patients. Symptoms of heart failure symptoms such as breathlessness, orthopnea, and fatigue may be present and, although they are more common in the adult population, they may be observed in HCM patients of all ages. These can vary from mild symptoms to New York Heart Association (NYHA) III/IV heart failure symptoms in around 15%–20% [165]. Diastolic dysfunction is the principal cause of heart failure, due to impaired left ventricular relaxation and a stiffened left ventricle impeding LV filling [165]. Ultimately, this reduces cardiac output, leading to pulmonary venous congestion and hence decreased exercise tolerance. Those patients with co-existent atrial fibrillation (AF) may have more severe symptoms due to the loss of the atrial "kick," which occurs in normal atrial systole and aids ventricular filling. The risk of stroke is 1% annually in those with chronic AF, and therefore, anticoagulation is advised [167].

Syncope occurs in around 15%–20% of those with HCM [168]. The underlying etiology is complex and not limited to those with LVOTO. Broadly speaking, it can be explained by 2 mechanisms: arrhythmic syncope and syncope as a result of hemodynamic instability. In terms of arrhythmic causes, AF is fairly common in HCM, and paroxysms with a fast ventricular response may result in clinical deterioration with syncope or pre-syncope. Myocardial scarring or fibrosis due to small vessel ischemia can result in an unstable electrical substrate, which may precipitate non-sustained VT and also result in syncope. From a hemodynamic perspective, syncope can be explained by 3 main mechanisms: LVOTO, an abnormal blood pressure response to exercise, and hypotension due to impaired left ventricular filling [168].

Clinical examination in the absence of LVOTO obstruction or heart failure may be normal. In those patients who do have a resting LVOTO and left ventricular gradient, a bifid pulse and apical ejection systolic murmur may be found. The presence of LVOTO is of significant clinical importance as these patients have a more malignant condition than those who do not, with a higher incidence of SCD, higher likelihood to develop NYHA III/IV heart failure, and higher risk of death due to heart failure and stroke [169].

Of the patients with HCM, 75%–95% have an abnormal ECG [170]. The most common ECG abnormalities seen in patients with HCM are voltage criteria for left ventricular hypertrophy and widespread Q-waves [171]. T-wave inversions are also commonly found. Variants of HCM, for example apical HCM, can often show quite profound ECG abnormalities with deep TWIs across the chest leads, particularly the left precordial leads (V_4–V_6; Figure 13A) [172, 173]. Although an ECG in individuals with athlete's heart may show isolated voltage, criteria for LVH, TWI, and Q-waves are rarer in the absence of pathology, although this is dependent on age and ethnicity (see section 3.3.1).

Echocardiographic findings are variable in patients with HCM, but the majority will have evidence of LVH. Classically, this is asymmetrical septal hypertrophy (Figure 13B), but other variants exist such as apical and concentric hypertrophy. Any of the following on 2D echocardiography fulfil the criteria for HCM: unexplained maximal wall thickness of >15 mm in any myocardial segment; septal/posterior wall thickness ratio of >1.3 in normotensive patients; septal/posterior wall thickness ratio of >1.5 in hypertensive patients [161].

Other investigations in HCM patients to identify the broader phenotypic features of the condition include CMRI, 24-hour Holter monitor, and treadmill testing. Cardiac MRI is used to give a better understanding of the structure of the heart and assess for evidence of scarring through the presence or absence of late-gadolinium enhancement (Figure 13C). Exercise treadmill testing and 24-hour Holter monitoring are used to risk-stratify patients. High-risk features include unheralded syncope, non-sustained ventricular tachycardia (NSVT), a family history of SCD, LVH of >30 mm, and an attenuated/reduced blood pressure to exercise (defined as a failure of systolic blood pressure to rise by >25 mm Hg with exercise) [41].

Treatment depends on the specific disease pattern and symptoms exhibited by the patient. Asymptomatic patients are given lifestyle advice regarding limitation of physical activity, ensuring that they keep well hydrated. All patients are followed up with annual surveillance echocardiography and risk stratification tests as described above. Those with symptoms such as angina or breathlessness secondary to diastolic dysfunction (i.e., heart failure with preserved ejection fraction) are treated with negatively inotropic agents such as a beta-blocker or verapamil [140, 165, 174, 175]. Beta-blockers should be used with caution in those with a resting bradycardia or severe conduction defects, and verapamil should be used with caution in those with bradycardia or a very high resting outflow tract gradient [175]. Some may not have amelioration of their symptoms with the initial agent and can be switched to a different one, but there does not appear to be benefit from combined therapy. In the end stages of HCM (also known as the "burnt-out" phase) where patients develop

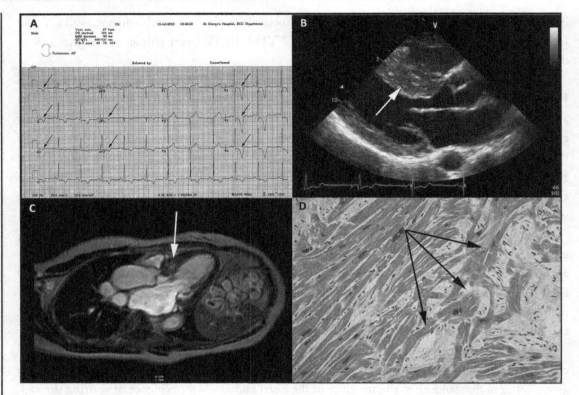

FIGURE 13: Clinical manifestations of HCM. A: ECG from a patient with apical HCM. Note the voltage criteria for LVH and deep TWIs (arrows); B: Transthoracic echocardiogram showing gross (2–3 cm) asymmetrical septal hypertrophy; C; CMRI showing LVH and late gadolinium enhancement (arrow); D: Histology from a patient with HCM showing myocyte disarray (arrows).

left ventricular systolic dysfunction, conventional heart failure treatment with diuretics, vasodilators, or digitalis maybe be appropriate [140, 165, 175–177]. Beta-blockers are the treatment of choice in those who develop AF or who have evidence of NSVT on 24-hour Holter monitoring. Patients should be counselled regarding the possibility of an implantable cardioverter–defibrillator (ICD) when high-risk features are present, which include the following:

1. A personal history for ventricular fibrillation (VF), sustained ventricular tachycardia (VT), or SCD events, including appropriate ICD therapy for ventricular tachyarrhythmias *(Class I indication)*
2. A family history for SCD events, including appropriate ICD therapy for ventricular tachyarrhythmias *(Class IIa indication)*

3. Unexplained syncope *(Class IIa indication)*
4. Documented NSVT, defined as 3 or more beats at greater than or equal to 120 bpm on ambulatory ECG monitoring *(Class IIa indication)*
5. Maximal LV wall thickness greater than or equal to 30 mm *(Class IIa indication)*

American guidelines recommend ICD implantation in those with ≥1 high-risk features, although in the United Kingdom, their used is reserved for patients with ≥2 high-risk features. However the decision regarding ICD must be individualized, taking into account the risk-benefit in each patient, with discussion of the pros and cons to allow an informed decision regarding implantation [175].

Patients with LVOTO are treated with negatively inotropic agents such as beta-blockers or verapamil. For those with ongoing symptoms despite maximal dosage, disopyramide may be an alternative or added into the treatment regimen [165, 175]. In severe cases, refractory to pharmacological therapy, surgical myotomy–myomectomy may be considered; in selected cases, alcohol septal ablation is an alternative.

4.1.1.3 Hypertrophic Cardiomyopathy in Sport. Many of the principles underpinning the management of HCM in the general population are applicable to athletes; however, there are several key aspects that need to be considered when assessing athletic individuals. Although the prevalence of HCM among trained athletes appears to be lower than in the general population (in the region of 0.07% [178]–0.09% [179]), the condition appears to be the commonest cause of SCD in US athletes, with deaths higher in athletes of African/Afro-Caribbean orign [35]. In their 10-year review of events during collegiate and high school sports, Mueller et al. [180] found all but 1 of the 56 subjects whose sudden death were attributed to HCM to be male. Maron et al. [10] found a similar male predominance in their review of competitive young athletes, with only 2 of the 48 cases of SCD in HCM being female. As mentioned above, some data have suggested an ethnic predilection, with more deaths observed in black athletes [35]. It is felt the high uptake of competitive sports in this cohort alone cannot explain this finding [10, 35]. Deaths appear to be commoner in intermittent "stop–start" sports such as soccer, basketball, and American football [10, 12, 35, 181]. The role of pre-participation screening for HCM in athletes, based on the European model of history, physical examination, and 12-lead ECG, has been found to be efficacious in preventing SCD through disqualification of affected individuals from competitive sports [37, 178].

As discussed in section 3.3.1, athletic training may produce ECG changes that overlap with those observed in patients with HCM. These physiological changes are more marked in black athletes, which is a phenomenon that has incited a lot of interest recently among sports cardiologists. Like ECG interpretation, the differentiation of LVH due to morphologically mild HCM from

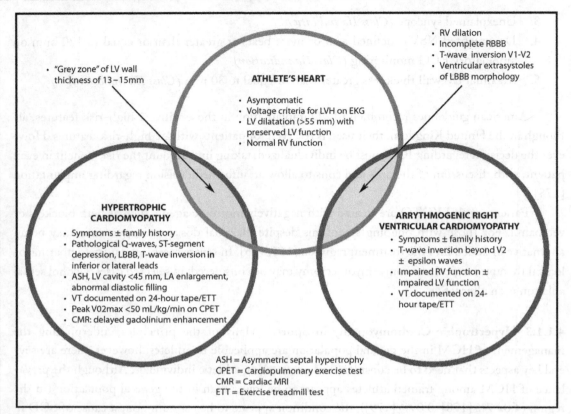

FIGURE 14: A practical guide for differentiating the physiological changes seen in "athlete's heart" from those of pathology in HCM and ARVC.

LVH secondary to athlete's heart can be particularly difficult in trained athletes. The differentiation between physiological and pathological LVH becomes crucial in these athletes when one considers the consequences of an erroneous diagnosis. Athletes with LVH in the range of 13–16 mm fall within the so-called "gray-zone" of hypertrophy (Figure 14). There are several clinical features that may help distinguish between athlete's heart and HCM when wall thicknesses fall into this gray zone. Features in the history that point toward a diagnosis of HCM include episodes of unheralded syncope, exertional chest pain, breathlessness, and palpitation. A detailed family history may reveal other individuals with HCM or a family history of sudden cardiac death. Abnormalities on the resting ECG suggestive of HCM rather than athlete's heart include pathological Q-waves, resting ST-segment depression, LBBB, and T-waves inversions, particularly in the lateral and/or inferior leads.

Echocardiography is invaluable in helping differentiate athlete's heart from HCM. The pattern of LVH in physiological remodelling is usually homogenous and eccentric in nature (i.e., associated with an enlarged LV chamber size); in addition, indices of diastolic function are normal. In contrast, HCM is associated with asymmetric and atypical patterns of LVH including septal, apical, and rarely posterior–lateral wall hypertrophy. Patients with HCM usually exhibit small LV cavity dimensions (<45 mm), and both pulsed-wave and tissue-Doppler indices of diastolic filling are impaired. Other features on echocardiography favouring HCM include a resting LV outflow tract gradient and systolic anterior motion of the mitral valve leaflet.

Several additional investigations may help in this diagnostic dilemma. A blunted blood pressure (BP) response to exercise and the presence of NSVT on Holter monitoring favor a diagnosis of HCM. Cardiac MRI may be crucial in detecting segmental LVH in regions that can sometimes be poorly visualised on echocardiography (such as the LV apex) and may also demonstrate delayed gadolinium enhancement indicative of myocardial fibrosis or scarring. In difficult cases, cardiopulmonary exercise testing (CPEX) can be useful given that patients with HCM invariably have a low peak oxygen consumption (<50 mL/kg/min) [182]. The mechanism behind this is a failure to augment stroke volume, due to a combination of LVH, myocyte ischemia, and fibrosis, which together impair diastolic filling; dynamic LVOTO also plays a role [183].

A summary and practical guide in differentiating athlete's heart from HCM is given in Figure 14. Genetic testing is currently of limited value in athletes, being time consuming, costly, and having a low negative predictive value. However, assessment of an athlete's first-degree relatives for phenotypic features of HCM can be extremely valuable. In rare cases, and where the athlete agrees, re-evaluation after a period of detraining may solve the dilemma given that cardiac adaptive changes regress as early as 6 weeks after the cessation of regular intensive exercise [184].

Although most athletes attempting to compete at elite level are selected out due to an inability to augment stroke volume, in some cases, those with morphologically mild HCM (particularly the apical variant) and a compliant LV, along with normal diastolic and systolic parameters, may be capable of high athletic attainment. Indeed, there have been several reports published in the literature describing athletes with HCM successfully participating in ultra-endurance events such as marathons [181, 185, 186].

As mentioned previously, pre-participation screening with 12-lead ECG is useful in identifying athletes with HCM and is discussed in more detail in section 9. Unfortunately, the diagnosis has major implications for any patient, but this is particularly true in the athlete. Those with a definite diagnosis of HCM with possible high-risk features are excluded from all competitive sports to minimize the risk of SCD. Those with a definite diagnosis but none of the high-risk features are allowed to participate in low-static, low-dynamic sports such as cricket, golf, or bowling (refer to Table 4 for classification of sport). Patients with a positive genotype but negative phenotype are difficult to manage: although they do not express the condition at presentation, they may develop the condition later on in life. The American College of Cardiology does not recommend exclusion from competitive sports on the basis of a positive phenotype [187]; however, the European recommendations are much more stringent, precluding the athlete from all but recreational sporting activities [188].

Below, we present 2 case studies, which highlight the challenges encountered in clinical practice when assessing athletes with phenotypic features of HCM.

4.1.1.4 Hypertrophic Cardiomyopathy Case Study 1.
The ECG of an asymptomatic 20-year-old regional club football player of West African ethnicity is shown in Figure 15 below that was

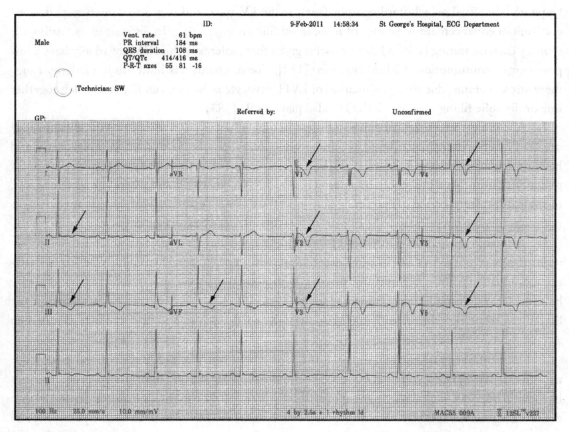

FIGURE 15: ECG of an asymptomatic 20-year-old regional club football player of West African ethnicity. Note the widespread TWIs across the anterior precordial leads V1–V4, extending into the inferolateral leads V5–V6, II, III, and aVL (arrows) and Sokolow–Lyon voltage criteria for left ventricular hypertrophy.

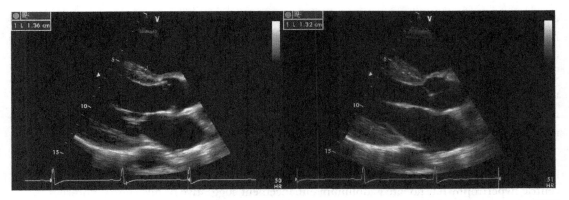

FIGURE 16: An echocardiogram from the West African athlete, which revealed mild concentric LVH of 13 mm.

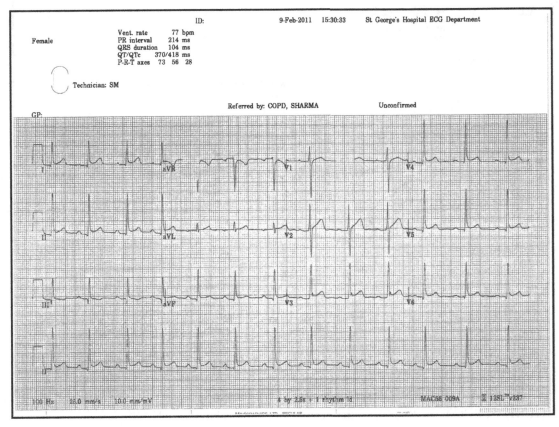

FIGURE 17: A normal ECG from the athlete's sister.

performed as part of a pre-participation cardiovascular evaluation. He exercised an average of 8–9 hours per week. Note the widespread TWIs across the anterior precordial leads V1–V4, extending into the inferolateral leads V5–V6, II, III, and aVL (arrows), along with Sokolow–Lyon voltage criteria for left ventricular hypertrophy. The athlete went on to have an echocardiogram, which revealed mild concentric LVH (Figure 16) of 13 mm, but no other phenotypic features of a cardiomyopathy. There was no family history of note, and an ECG performed in his sister (Figure 17) was normal, with no phenotypic features of HCM.

Further investigations including an exercise stress test and 24-hour Holter monitoring were performed and found to be normal. He went on to have a CMRI scan, which confirmed mild concentric hypertrophy but no late gadolinium enhancement suggestive of myocardial fibrosis.

On routine follow-up 6 months later, the athlete mentioned that he had injured his ankle and had therefore been unable to play football. A repeat ECG was performed, and is shown in Figure 18

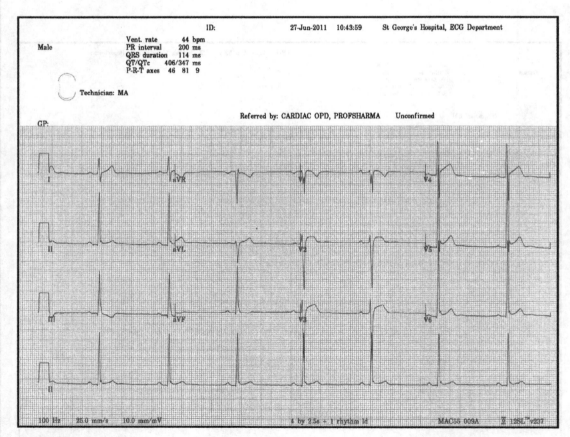

FIGURE 18: ECG from the athlete after detraining. Note resolution of all TWIs.

below, which shows complete resolution of the TWIs, leaving changes of concave ST-segment elevation, voltage criteria for LVH, and early repolarization, all of which can be normal for an athlete. A repeat echocardiogram was performed, which again showed mild concentric LVH at 12 mm.

4.1.1.5 Discussion to Case Study 1. This case highlights a number of important points and difficulties that can be encountered with regard to black athletes who are known to develop more TWIs (including deep TWIs) and LVH in comparison to white athletes. As discussed in sections 3 and 7, from our experience TWIs confined to V1–V4, particularly when preceded by concave ST-segment elevation, may be considered benign as long at the athlete is asymptomatic, has no family history of note, and no broader phenotypic features of a cardiomyopathy. In addition, these changes have been seen to resolve with detraining.

However, this example demonstrates TWIs extending into the inferior and lateral leads. Studies in both white and black athletic populations have demonstrated an association (albeit in a small minority) between TWIs in the lateral ECG leads and subsequent development of a cardiomyopathy [54, 101]. In addition, this athlete had mild LVH that although easily explained by his degree of athletic training (particularly in the context of his ethnicity) does raise the possibility of morphologically mild HCM.

The resolution of T-wave inversions with detraining suggests that these electrical manifestations may represent an electrical response to physiological cardiac remodelling. However, it may be argued that the development of T-wave inversions may represent the subclinical expression of a cardiomyopathy whose phenotype has been exposed by intense exercise. In this case, the athlete was asymptomatic and had no other broader phenotypic features of HCM and the ECG of a first-degree relative was normal. His pattern of LVH on echocardiography was homogenous and did diminish with detraining. Cardiopulmonary exercise testing would have been helpful, given that patients with HCM invariably have a low peak oxygen consumption (<50 mL/kg/min; see Figure 14) [331]. An ECG and echocardiography on each of his parents would also have been extremely useful.

In the case of this athlete we adopted a pragmatic approach and allowed him to continue playing but on the proviso that he have regular follow-up with annual ECG and echocardiography; should he develop any symptoms, he was advised to seek medical attention to be reassessed immediately.

4.1.1.6 Hypertrophic Cardiomyopathy Case Study 2. A 16-year-old professional footballer of mixed European and West African descent attended clinic after a pre-participation screening event revealed an abnormal ECG. He exercised for over 30 hours per week and was on the verge of being signed by a top European football club. His ECG, shown in Figure 19 below, revealed deep TWIs in leads III and V4–V6 (arrows) and biphasic T-waves in II and aVF (arrowheads). There was a family history of sudden death, but the cause of this had never been established. Given his elite status and this crucial time in his career, he was not willing to detrain. For comparison, the ECG from a patient with known apical HCM (phenotypic features + gene positive) is shown in Figure 20 below, with deep TWIs in a similar distribution (arrows).

The athlete went on to have an echocardiogram, which showed no clear evidence to suggest HCM. Exercise testing and a 24-hour Holter monitoring were also normal, as was CMRI and his mother's ECG. In particular, the LV apex was of normal size on his echocardiogram and CMRI.

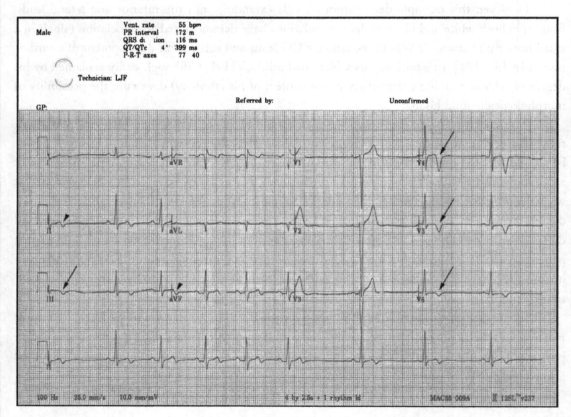

FIGURE 19: ECG from a 16-year-old professional footballer of mixed European and West African descent revealed deep TWIs in leads III and V4–V6 (arrows) and biphasic T-waves in II and aVF (arrowheads).

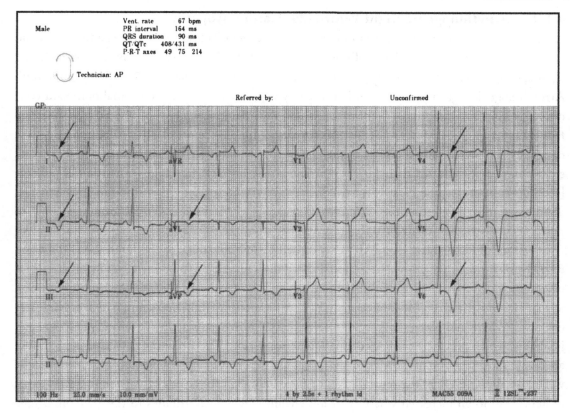

FIGURE 20: ECG from a patient with known apical HCM. Note deep TWIs in the infero-lateral leads.

4.1.1.7 Discussion to Case Study 2. This case is similar to Case Study 1, but we do not have the benefit of an ECG after detraining. Often in reality, top professional athletes are unwilling to detrain; however, note the similarities in TWI distribution to the ECG from a patient with apical HCM (Figure 20). In this case, the clinician is faced with a slightly difficult situation—the only evidence of an abnormality is the isolated TWIs in the lateral leads; do these represent athletic training or the early manifestations of HCM? It is recognized that TWIs in the lateral leads in particular are common in HCM but are also present in 4% of black male athletes who do not have any features of HCM on imaging studies.

In this case, the athlete was once again allowed to continue playing, given that no broader phenotypic features of HCM were present, on the strict instruction that, should any symptoms develop, he be reassessed. Careful monitoring with an annual ECG and echocardiogram were instituted.

4.1.2   Arrhythmogenic Right Ventricular Cardiomyopathy

Arrhythmogenic right ventricular cardiomyopathy is a rare condition that affects adolescents and young adults. The exact prevalence of this condition is difficult to determine given its variable and age-dependent penetrance but is estimated to be between 1 in 1000 and 1 in 5000 [139, 189]. Arrhythmogenic right ventricular cardiomyopathy is predominantly an autosomal dominant condition, although an autosomal recessive variant is described and approximately 50% of cases are familial [190]. There appears to be male to female predominance of 3:1 [191]. The condition was classically considered to be a disease of the right ventricle; however, recent data from post-mortem studies have demonstrated that the left ventricle is also involved in a high proportion of patients [192]. In light of this, the Heart Rhythm Society and the European Heart Rhythm Society issued a joint statement suggesting the condition be renamed as "Arrhythmogenic Cardiomyopathy" to reflect its biventricular nature [193].

At an ultrastructural level, ARVC is caused by abnormalities of the intercalated discs and related structures. In the normal heart, intercalated discs function to bind myocytes together and facilitate their controlled and synchronised contraction simultaneously, allowing the heart to function as a solid organ. Between the intercalated discs are desmosomes (glycoproteins, which provide mechanical strength and play an important role in the anchoring and stability of ion channels) and gap junctions, which provide electrical conduction between cells. The structural defects seen in ARVC are due to the disruption of the intercellular communication and cell proliferation/differentiation, as well as an abnormality in the transmission of force between cells [139]. Mutations in desmosomal proteins lead to an increased volatility of ion channels, which contributes to the arrhythmogenic nature of the condition [194]. Three types of proteins make up the desmosomes: desmosomal cadherins, armadillo proteins (plakoglobin and plakophilin), and plakins (Figure 21).

Pathologically, ARCV is characterized by fibrofatty replacement of the ventricular wall from the epicardium to the endocardium [195] (Figure 22E, black arrow). This fibrofatty replacement eventually interferes with intraventricular conduction, which can precipitate a re-entry phenomenon and hence ventricular arrhythmias [195]. Fibrofatty replacement also weakens the myocardium, predisposing to dilatation and aneurysm formation [196]. It is important to note that fat replacement of the myocardium (adipositas cordis) alone is not pathognomonic of ARVC, as this may be present in normal hearts in the absence of underlying pathology; it is myocyte degeneration and replacement with fibrosis (Figure 22E, yellow arrow) that provide the classical hallmark of this condition and that need to be present to establish a firm diagnosis [197]. It is thought the fatty infiltration is secondary to the transdifferentiation of cardiomyocytes to adipocytes [198] due to nuclear translocation of plakoglobin secondary to the suppression of desmoplakin.

The initial understanding of the genetic basis of this condition came from the discovery of the 2 rare autosomal recessive cardio-cutaneous conditions associated with ARVC, Naxos dis-

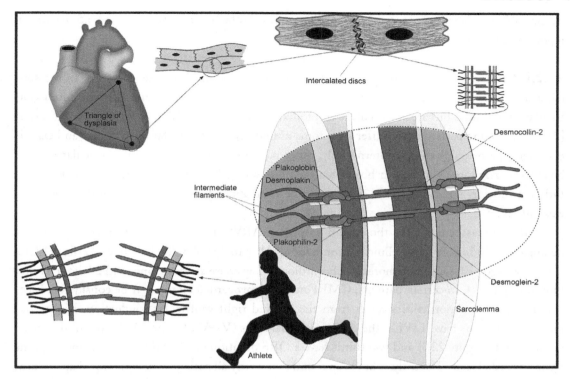

FIGURE 21: Schematic representation of pathological processes causing ARVC and the proposed mechanism by which exercise causes RV stress and pathology.

ease, and Carvajal syndrome. The first mutation discovered was a 2-bp deletion in JUP-encoded plakoglobin seen in Naxos disease [199]. Carvajal syndrome was later discovered and found to be due to a point mutation in DSP-encoded desmoplakin [200]. Later, mutations in desmoplakin [201], plakophilin-2 [202], desmoglein-2 [203] and desmocollin-2 [204] were discovered. It is thought that PKP2 mutations make up approximately 45% of those with a known genetic cause that meet the task force criteria [205]; however, this is thought to represent a much higher proportion in familial cases, maybe up to 70% [189]. DSP mutations are seen in 6%–16% of ARVC cases [201, 206]. Among those without a PKP2 or DSP mutation, 5%–10% of cases have mutations in DSG2 or DSC2 [203, 204]. Non-desmosomal protein mutations have also been discovered. A mutation in the transmembrane protein TMEM43 was discovered among related family members in Newfoundland in Canadia [207]. Mutations in the transforming growth factor B protein, which is important for embryogenesis and cell differentiation and encoded by the TGFB3 gene, have been discovered in a single cohort [208]. Furthermore, patients with mutation in the RyR2 gene (responsible for catecholaminergic polymorphic ventricular tachycardia, see section 4.4.5) have been reported to have an ARVC phenotype [209].

4.1.2.1 Clinical Features and Management.   Four main clinical pictures are seen. In the subclinical or "concealed" phase, patients may have only subtle structural abnormalities or minor ventricular arrhythmias but are essentially asymptomatic although may still present with cardiac arrest [210, 211]. Some patients may suffer symptoms of underlying ventricular arrhythmias and experience syncope or palpitations. Alternatively, other patients present with right ventricular or even biventricular failure. Those with biventricular failure are at high risk of mural thrombus due to turbulent flow in aneurysmal areas of the heart or atrial appendages, particularly in those with co-existent AF.

Routine assessment for those with suspected ARVC includes a full history (including a detailed family history) and clinical examination. All patients should then have 12-lead ECG, signal-averaged ECG, a transthoracic echocardiogram, exercise stress testing, and 24-hour Holter monitoring [212]. In certain patients, CMRI and endomyocardial biopsy may be considered. Electrocardiographic abnormalities result from the delayed right ventricular conduction. Eighty-five percent of patients have TWI in the right precordial leads (V_1–V_3; Figure 22A), one third have an epsilon wave (Figure 22B) and two thirds have a QRS duration of >110 milliseconds (msec) in the right precordial leads. Recently, it has been found that nearly all patients have prolonged upstroke of the S wave (>55 msec) in the right precordial leads, and this is now considered a diagnostic marker on ECG [213]. Other less specific findings include a right bundle branch block pattern. Signal-averaged ECG may show fragmented low amplitude late potentials at the end of the QRS complex [190]. Twenty-four-hour Holter monitors are important in terms of demonstrating the broader phenotype of the disorder and identifying those with episodes of NSVT. Patients may experience any ventricular arrhythmia, from premature ventricular complexes to sustained VT and ventricular fibrillation. In those with ventricular ectopics or episodes of NSVT, the morphology of the QRS complexes may help locate the site of origin. Ventricular ectopics or NSVT with LBBB morphology and inferior axis usually originate from the right ventricular outflow tract, whereas LBBB with superior axis morphology is likely to originate from the right ventricular inferior wall [190]. However these findings are not specific and may be present in idopathic right ventricular outflow tract tachycardia.

Standard transthoracic echocardiography can give sufficient delineation of the right ventricle. Findings consistent with ARVC include global RV dilation with or without RV systolic dysfunction or LV involvement, segmental RV dilation with or without dyskinesia, and RV wall motion abnormalities (Figure 22C) [196, 214, 215]. In severe cases of ARVC, CMRI adds little extra information; however, it is superior to echocardiography at detecting abnormalities in those with the concealed

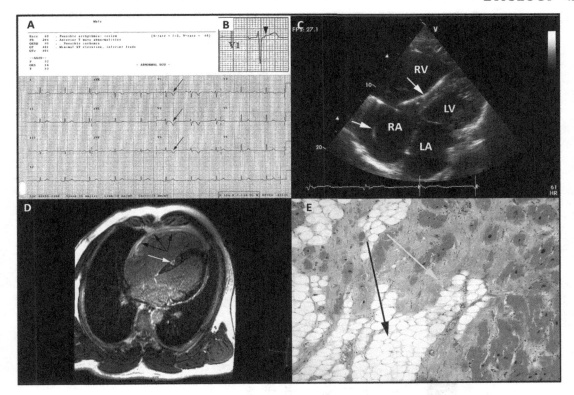

FIGURE 22: Clinical features of ARVC. A: ECG demonstrating TWIs in the right precordial leads V1–V3; B: epsilon wave signifying delayed depolarisation of the right ventricle; C: Echocardiogram demonstrating gross RV dilatation. Note the presence of an ICD lead (arrows); D: CMRI scan demonstrating biventricular involvement with late gadolinium enhancement in the RV (black arrows) and LV septum (white arrow); E: Histology showing both fatty infiltration (black arrow) and replacement fibrosis (yellow arrow).

phase of the condition and when screening family members and allows further tissue characterisation and detection of late gadolinium enhancement indicative of intramyocardial fibrosis (Figure 22D). Rarely, endomyocardial biopsy is performed when ambiguity of the diagnosis remains. Biopsy is particularly useful when distinguishing between ARVC and inflammatory/autoimmune conditions such as sarcoidosis and myocarditis [190]. The decision to perform this, however, should not be taken lightly given that biopsies are taken from the RV free wall, which is an extremely thin structure at risk of perforation with potentially catastrophic consequences. Ultimately, diagnosis is based upon fulfilling the task force criteria (Table 6). Patients are said to fulfil these criteria in either the following circumstances: 2 major criteria, 1 major and 2 minor criteria, or 4 minor criteria (from

TABLE 6: Comparison of Original and Revised Task Force Criteria. Taken from Marcus et al. [222] with permission.

ORIGINAL TASK FORCE CRITERIA	REVISED TASK FORCE CRITERIA
I. Global or regional dysfunction and structural alterations*	
Major	
• Severe dilatation and reduction of RV ejection fraction with no (or only mild) LV impairment • Localized RV aneurysms (akinetic or dyskinetic areas with diastolic bulging) • Severe segmental dilatation of the RV	**By 2D echo:** • Regional RV akinesia, dyskinesia, or aneurysm *and* 1 of the following (end diastole): — PLAX RVOT ≥32 mm (corrected for body size [PLAX/BSA] ≥19 mm/m^2) — PSAX RVOT ≥36 mm (corrected for body size [PSAX/BSA] ≥21 mm/m^2) — or fractional area change ≤33% **By MRI:** • Regional RV akinesia or dyskinesia or dyssynchronous RV contraction *and* 1 of the following: — Ratio of RV end-diastolic volume to BSA ≥110 mL/m^2 (male) or ≥100 mL/m^2 (female) — *or* RV ejection fraction ≤40% **By RV angiography;** • Regional RV akinesia, dyskinesia, or aneurysm

Minor	

	• Mild global RV dilatation and/or ejection fraction reduction with normal LV
	• Mild segmental dilatation of the RV
	• Regional RV hypokinesia
		By 2D echo:
		• Regional RV akinesia or dyskinesia
		• *and* 1 of the following (end diastole):
		  — PLAX RVOT ≥29 to <32 mm (corrected for body size [PLAX/BSA] ≥16 to <19 mm/m^2)
		  — PSAX RVOT ≥32 to <36 mm (corrected for body size [PSAX/BSA] ≥18 to <21 mm/m^2)
		  — *or* fractional area change >33% to ≤40%
		
		By MRI:
		• Regional RV akinesia or dyskinesia or dyssynchronous RV contraction
		• *and* 1 of the following:
		  — Ratio of RV end-diastolic volume to BSA ≥100 to <110 mL/m^2 (male) or ≥90 to <100 mL/m^2 (female)
		  — *or* RV ejection fraction >40% to ≤45%

TABLE 6: *(continued)*

ORIGINAL TASK FORCE CRITERIA	REVISED TASK FORCE CRITERIA
II. Tissue characterization of wall	
Major	
• Fibrofatty replacement of myocardium on endomyocardial biopsy	• Residual myocytes <60% by morphometric analysis (or <50% if estimated), with fibrous replacement of the RV free wall myocardium in ≥1 sample, with or without fatty replacement of tissue on endomyocardial biopsy
Minor	
	• Residual myocytes 60% to 75% by morphometric analysis (or 50% to 65% if estimated), with fibrous replacement of the RV free wall myocardium in ≥1 sample, with or without fatty replacement of tissue on endomyocardial biopsy
III. Repolarization abnormalities	
Major	
	• Inverted T waves in right precordial leads (V_1, V_2, and V_3) or beyond in individuals >14 years of age (in the absence of complete right bundle–branch block QRS ≥120 ms)

Minor	• Inverted T waves in right precordial leads (V_2 and V_3) (people age >12 years, in absence of right bundle–branch block)	• Inverted T waves in leads V_1 and V_2 in individuals >14 years of age (in the absence of complete right bundle–branch block) or in V_4, V_5, or V_6 • Inverted T waves in leads V_1, V_2, V_3, and V_4 in individuals >14 years of age in the presence of complete right bundle–branch block

IV. Depolarization/conduction abnormalities

Major	• Epsilon waves or localized prolongation (>110 ms) of the QRS complex in right precordial leads (V_1 to V_3)	• Epsilon wave (reproducible low-amplitude signals between end of QRS complex to onset of the T wave) in the right precordial leads (V_1 to V_3)
Minor	• Late potentials (SAECG)	• Late potentials by SAECG in ≥1 of 3 parameters in the absence of a QRS duration of ≥110 ms on the standard ECG • Filtered QRS duration (fQRS) ≥114 ms • Duration of terminal QRS <40 µV (low-amplitude signal duration) ≥38 ms • Root-mean-square voltage of terminal 40 ms ≤20 µV • Terminal activation duration of QRS ≥55 ms measured from the nadir of the S wave to the end of the QRS, including R′, in V_1, V_2, or V_3, in the absence of complete right bundle–branch block

TABLE 6: (*continued*)

	ORIGINAL TASK FORCE CRITERIA	REVISED TASK FORCE CRITERIA
V. Arrhythmias		
Major		• Nonsustained or sustained ventricular tachycardia of left bundle-branch morphology with superior axis (negative or indeterminate QRS in leads II, III, and aVF and positive in lead aVL)
Minor	• Left bundle-branch block-type ventricular tachycardia (sustained and nonsustained) (EGG, Holter, exercise) • Frequent ventricular extrasystoles (>1000 per 24 hours) (Holter)	• Nonsustained or sustained ventricular tachycardia of RV outflow configuration, left bundle-branch block morphology with inferior axis (positive QRS in leads II, III, and aVF and negative in lead aVL) or of unknown axis • >500 ventricular extrasystoles per 24 hours (Holter)
VI. Family history		
Major	• Familial disease confirmed at necropsy or surgery	• ARVC/D confirmed in a first-degree relative who meets current Task Force criteria • ARVC/D confirmed pathologically at autopsy or surgery in a first-degree relative

	• Identification of a pathogenic mutation[†] categorized as associated or probably associated with ARVC/D in the patient under evaluation
Minor	• Family history of premature sudden death (<35 years of age) due to suspected ARVC/D • Familial history (clinical diagnosis based on present criteria)	• History of ARVC/D in a first-degree relative in whom it is not possible or practical to determine whether the family member meets current Task Force criteria • Premature sudden death (<35 years of age) due to suspected ARVC/D in a first-degree relative • ARVC/D confirmed pathologically or by current Task Force Criteria in second-degree relative

PLAX indicates parasternal long-axis view; RVOT, RV outflow tract; BSA, body surface area; PSAX, parasternal short-axis view; aVF, augmented voltage unipolar left foot lead; and aVL, augmented voltage unipolar left arm lead.

Diagnostic terminology for original criteria: This diagnosis is fulfilled by the presence of 2 major, or 1 major plus 2 minor criteria or 4 minor criteria from different groups. Diagnostic terminology for revised criteria: definite diagnosis: 2 major or 1 major and 2 minor criteria or 4 minor from different categories; borderline: 1 major and 1 minor or 3 minor criteria from different categories; possible: 1 major or 2 minor criteria from different categories.

*Hypokinesis is not included in this or subsequent definitions of RV regional wall motion abnormalities for the proposed modified criteria.

[†]A pathogenic mutation is a DNA alteration associated with ARVC/D that alters or is expected to alter the encoded protein, is unobserved or rare in a large non-ARVC/D control population, and either alters or is predicted to alter the structure of function or the protein or has demonstrated linkage to the disease phenotype in a conclusive pedigree.

different groups) [216]. The task force criteria are used to diagnose probands, but modifications have been suggested when diagnosing asymptomatic patients through family screening.

Management of patients with ARVC depends largely upon their symptoms and the presence or absence of high-risk features. It is thought that manifestations of this condition are precipitated by sympathetic stimulation, and hence, those with symptoms are put on beta-blockers as first-line therapy. Amiodarone can be added in for those whose symptoms do not resolve with beta-blockers alone [217]. Some would favor sotalol; however, cautious monitoring of the corrected-QT interval is required in these cases [218]. Most of the evidence for the use of pharmacological therapy is anecdotal given that there is a lack of trial data on the efficacy of these drugs. It is unknown whether beta-blockers alter the natural progression of this condition, and hence, pharmacological therapy is not indicated in asymptomatic individuals. In very symptomatic patients with ectopy or arrhythmias refractory to pharmacological therapy, catheter ablation may be considered. Its use is limited however as, although initial results are good, there is a 90% recurrence in symptomatic hemodynamically stable ventricular arrhythmias at a 3-year follow-up [219]. Heart failure in ARVC is treated with conventional therapy, and anticoagulation is recommended in those with AF, aneurysms, or marked ventricular dilation [217]. In a young patient with disabling symptoms and no improvement with conventional therapy, heart transplantation may be considered.

Decisions about who should receive an ICD are complex. There is little doubt that survivors of cardiac arrest should have an ICD fitted, but the decision is less clear cut in other cases. Those who are asymptomatic or have minimal symptoms, with no high-risk features, should not have an ICD inserted as they have a favorable long-term prognosis and the risks of complications from insertion are not insignificant [220]. In patients with syncope or compromised ventricular arrhythmias not responsive to pharmacological therapy, an ICD may be considered [220]. In these cases, the anti-tachycardia pacing setting is often utilized to terminate symptomatic non-fatal arrhythmias and hence reduces the need for high-dose pharmacological therapy. Current indications for insertion of an ICD according to the latest guidelines [221] and research [220] include the following:

1. A personal history for aborted SCD *(Class I indication)*
2. A personal history of sustained VT or VF *(Class I indication)*
3. Moderate RV dysfunction and arrhythmia
4. Those with extensive disease, including LV involvement *(Class IIa indication)*
5. Those with 1 or more affected family member with SCD *(Class IIa indication)*

4.1.2.2 Genotype–Phenotype Correlations.   There are various factors that seem to influence the severity and nature of the phenotype in ARVC. In general, missense mutations afford a worse prognosis than mutations leading to truncated proteins [223], and those with multiple mutations have significantly more manifestations of the disease [224]. Previous evidence suggested a high incidence of left ventricular involvement and sudden death in those with desmoplakin mutations [225], whereas a more recent, larger study found more profound LV involvement in desmoglein-2 mutations [223]. It has been reported that patients with plakophilin mutations exert a more severe phenotype, with a high incidence of life-threatening arrhythmias [226].

Genetic testing is can be quite difficult in these patients. Although desmosomal mutations will be present in around 50% of those who meet the task force criteria, it is felt that single desmosomal mutations maybe present in up to 16% of healthy volunteers [227]. There also appears to be an unclear pattern of penetrance even in the more common mutations. Genetic testing in parents of probands has found non-penetrant mutations, which are felt to be responsible for the phenotype when passed to their offspring [223]. Furthermore, this may lead to a diagnostic conundrum when screening siblings or offspring of probands who at the time of screening do not exhibit the clinical phenotype and leads to a dilemma as to how long these relatives should be followed up [139].

4.1.2.3 Arrhythmogenic Right Ventricular Cardiomyopathy in Sport.

Although ARVC is rare, it is not an uncommon cause of SCD in the athlete, being the most common cause in athletes in Northern Italy [11]. It is estimated that there is a 5-fold increase in sudden death in patients with ARVC during sporting activities. It is hypothesized that this is due to the hemodynamic changes that occur during exercise, with volume overload leading to stretching of the RV and sympathetic stimulation [32]. Moreover, excessive long-term mechanical stress aggravates myocardial lesions and accelerates disease progression (Figure 21). As discussed in section 3.3.2, it is now well established that, as with the LV, the RV also undergoes physiological dilatation with training [124], with up to 50% of athletes exhibiting RV and/or RV outflow tract dilatation and 6% fulfilling recent ARCV Task Force criteria [220]. Some recent studies, such as that by Heidbüchel et al. [228], have observed complex RV arrhythmias in endurance athletes, with a high proportion exhibiting concomitant RV structural abnormalities on angiography and CMRI and approximately 90% fulfilling diagnostic criteria for ARVC. Recently and somewhat controversially, it has even been postulated that long-term bouts of intensive exercise are associated with RV dysfunction and structural abnormalities, which may mimic those observed in ARVC, leading to an "acquired" form of the condition [229].

Arrhythmogenic right ventricular cardiomyopathy is therefore a particularly challenging diagnosis to make in the case of a highly trained athlete. Given that diagnosis is based on a combination of clinical features and several investigations (Table 6), this makes screening utilizing the 12-lead ECG comparatively much less useful than it is for HCM. The natural history of ARVC involves an early "concealed phase," during which the only manifestation of the condition include subtle electrical abnormalities on the right ventricular ECG leads (e.g., minor TWIs), mild RV dilatation, and ventricular ectopics of right ventricular outflow tract (RVOT) origin, all of which overlap with physiological adaptation of the right ventricle to exercise [230].

Figure 14 provides a practical guide to differentiating athlete's heart from ARVC. The presence of epsilon waves or late potentials on the signal-averaged ECG, NSVT of LBBB morphology on Holter monitoring, and regional wall motion abnormalities on either echocardiography or CMRI all favor a diagnosis of ARVC. As with HCM, the value of a thorough personal and family history cannot be underestimated, with symptoms (e.g., unheralded syncope, palpitation) and positive family history favoring a diagnosis of ARVC. Athletes with a definite diagnosis are advised to refrain from physical activity and are disqualified from all competitive sports [188].

We now present a case study highlighting the difficulties that may be encountered in clinical practice when assessing athletes with clinical features of ARVC.

4.1.2.4 Arrhythmogenic Right Ventricular Cardiomyopathy Case Study.   A 29-year-old Caucasian male professional Rugby player attended a routine pre-participation screening, where a 12-lead ECG and echocardiogram were performed. He was asymptomatic apart from occasional short-lived palpitation, and there was no family history of cardiac disorders or sudden death. His ECG revealed deep TWIs, confined to the right precordial leads V1–V3, as shown in Figure 23 below (arrows).

The echocardiogram revealed normal biventricular size and systolic function, with no evidence of right ventricular or outflow tract dilatation, no right ventricular regional wall motion abnormalities, and no aneurysmal dilatation to suggest a diagnosis of ARVC. For comparison, an ECG from a patient with known ARVC (symptoms, phenotypic features, and gene positive) is shown in Figure 24. Note the similar morphologies and distribution of the TWIs seen in the athlete's ECG in Figure 23.

A signal-averaged ECG was normal. The athlete went on to perform well on an exercise tolerance test, with no arrhythmias triggered. A 24-hour Holter monitor revealed occasional (<500) ventricular ectopics of LBBB morphology and superior axis.

4.1.2.5 Discussion to Case Study.   This case illustrates the overlap in ECG phenotype that can exist between athletic training and ARVC—note the similarities in the 2 ECGs. The ECG in the patient with ARVC also reveals an epsilon wave in V2, which if present in an athletic individual with inverted T-waves beyond V2 would be highly suggestive of cardiac pathology. In black athletes, such a pattern would be considered in keeping with athletic training; however, TWIs beyond V2 are extremely rare in Caucasian male athletes and should rigger comprehensive evaluation for an underlying cardiomyopathy. In this case, given that the TWIs were limited to the right precordial leads,

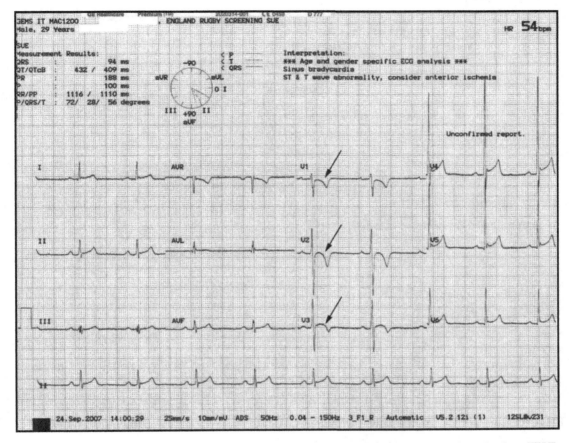

FIGURE 23: ECG from a 29-year-old Caucasian male professional Rugby player. Note the deep TWIs, confined to the right precordial leads V1–V3.

ARVC would be the main differential diagnosis, and therefore, assessment with signal-averaged ECG, echocardiography, exercise ECG, and Holter monitoring in the first instance is appropriate. In many cases, CMRI is also warranted.

When the results of this athlete were analyzed in combination, he met only 1 minor Task Force criteria for ARVC (see Table 6). He was therefore cleared for competition, but given the a diagnosis of ARVC can be extremely difficult to make (particularly in the early concealed phase of the condition) and that its natural history has not yet been fully determined, regular follow-up with yearly ECG and echocardiography was recommended. In addition, he was advised to return, should he develop any worrying symptoms including breathlessness, palpitation, or syncope.

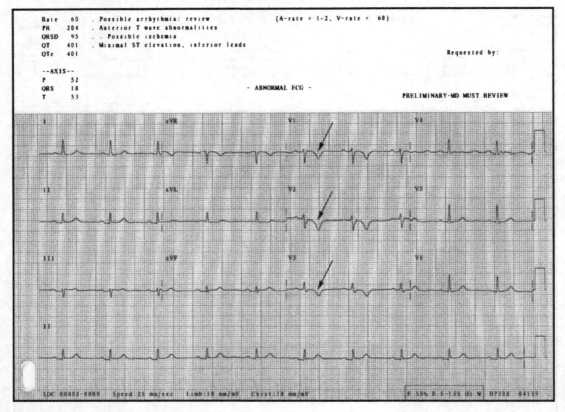

Rate 60 . Possible arrhythmia: review   (A-rate = 1-2, V-rate = 60)
PR 204 . Anterior T wave abnormalities
QRSD 95 . . Possible ischemia
QT 401 . Minimal ST elevation, inferior leads
QTc 401

Requested by:

--AXIS--
P 52
QRS 18   - ABNORMAL ECG -
T 53   PRELIMINARY-MD MUST REVIEW

LOC 00000-0000   Speed:25 mm/sec   Limb:10 mm/mV   Chest:10 mm/mV   N 50~ 0.5-150 Hz W   HP708 04139

FIGURE 24: An ECG from a patient with known ARVC. Note the similar distribution of TWIs.

4.1.3 Dilated Cardiomyopathy

The presence of left ventricular dilatation and systolic dysfunction are the hallmark of DCM. It has a prevalence of 1 in 2500 in the general population [231]. When all the detectable causes of DCM are excluded, the etiology remains uncertain in approximately 35%. This group of patients are described as having "idiopathic" DCM. Much research has been done into ascertaining the underlying cause in this group, with over 30 different genetic mutations being identified so far. It is felt, however, that this is only the tip of the iceberg [232] since it is estimated that, to date, only a third of all genetic mutations have been identified [233]. Broadly speaking, idiopathic DCM is classified as familial or sporadic. Strict criteria have been devised in defining familial DCM, with at least 2 closely related family members needing to be affected before a diagnosis of familial DCM can be made [234]. It is thought the vast majority of familial mutations are "private," that is, unique to that family and hence are not seen in multiple families [139]. An Italian study found that among those with familial DCM, more than half were autosomal dominant, 16% were autosomal recessive, 10% were X-linked, and 8% were due to mitochondrial disorders [235]. Screening of relatives of those with idiopathic DCM will find echocardiographic features of the condition in over one third of asymptomatic relatives [236–238].

The gene mutations seen in DCM, encode for a variety of proteins, which are responsible for myocyte structure and normal functioning. These include sarcomeric and related proteins, proteins involved in ion channel function, cytoskeletal proteins, nuclear envelope proteins, and those involved in mitochondrial functioning (Figure 25) [239]. It can be noted that these mutations share considerable overlap with those responsible for other inherited cardiac diseases such as HCM, ARVC, and congenital long-QT syndrome (LQTS). Mutations of LMNA gene are the most common genetic abnormality seen in patients with familial DCM, with a prevalence of 5%–8% in these patients. This gene normally encodes for nuclear lamins A and C, which are proteins critical in maintaining the cytoskeleton and structural integrity. The phenotypic expression of these mutations is very heterogeneous, with those affected having diverse disease spectrum. Dilated

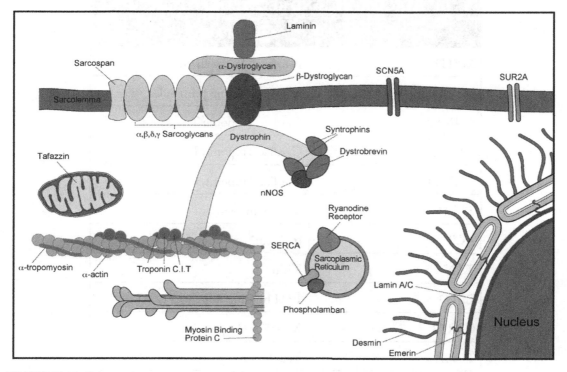

FIGURE 25: Schematic representation of the mutations involved in DCM.

cardiomyopathy with heart failure, DCM with predominantly arrhythmogenic disease, partial lipodystrophy, Emery–Dreifuss muscular dystrophy, and Hutchinso–Gilford progeria syndrome are all seen in patients with these mutations [139]. Of the remaining cases of DCM with a known mutation, the more common defects include those encoding the myosin heavy chain (MYH7), myosin-binding protein C (MYBPC3), troponin (TNNT, TNNC2), actin (ACTC1, ACTN2), and dystrophin (DMD) genes and SCN5A mutations [158, 240–245]. Rarer mutations causing DCM include those encoding for desmin, tafazzin, δ-sarcoglycan, phospholamban and the regulatory subunit of the ATP-sensitive cardiac K$^+$ channel (SUR2A; Table 7) [246–252]. If there is a predominance of conduction disease and atrial arrhythmias, then an LMNA mutation is more likely.

Alternatively, if familial DCM appears to follow an X-linked pattern of inheritance, then a dystrophin gene mutation may be more likely [158]. Other rare mutations are outlined in Table 7.

TABLE 7: Genetic mutations associated with dilated cardiomyopathy.

DILATED CARDIOMYOPATHY PROTEINS	
SARCOMERIC AND RELATED PROTEINS	
GENE	PROTEIN
MYH7	β-myosin heavy chain
MYBPC3	Myosin-binding protein C
TNNT2	Cardiac Troponin T
TNNI3	Cardiac Troponin I
TNNC1	Cardiac Troponin C
TPM1	α-tropomyosin
ACTC	α-Actin
TTN	Titin
LBD3	LIM binding domain 3 (ZASP)
CSRP3	Muscle LIM protein
TCAP	Telethonin
VLC	Vinculin/Metavinculin

TABLE 7: (*continued*)	
CYTOSKELETAL PROTEINS	
DMD	Dystrophin
SGCD	δ-sarcoglycan
INTERMEDIATE FILAMENTS	
	Desmin
NUCLEAR ENVELOPE PROTEINS	
LMNA	Lamin A/C
MITOCHONDRIAL DISORDERS	
G4.5	Tafazzin
CHANNELOPATHIES	
Gene	Channel
SCN5A	Nav1.5
SUR2A	ATP-sensitive potassium channel
PLN	Phospholamban

4.1.3.1 Clinical Features and Management.   Although they have a different genetic basis, familial and sporadic DCMs are clinically indistinct. Patients with either condition exhibit non-specific ECG changes, ventricular dilatation, systolic dysfunction, clinical heart failure, and atrial and ventricular arrhythmias. Further complications, including stroke and SCD, are seen in both conditions. Rarely, patients present with systemic thromboembolism or sudden death as their first presentation; however, most patients typically present with symptoms of heart failure such as breathlessness or fatigue. These symptoms are often insidious, but intercurrent illnesses may precipitate a decompensation, resulting in the patient presenting to a physician. Clinical examination may reveal features in keeping with heart failure, but it is important to look for stigmata of other conditions, which cause dilated cardiomyopathy such as valvular heart disease, infiltrative disorders (e.g., amylodosis, sarcoidosis, and hemochromatosis), endocrine disorders (thyroid disease, pheochromocytoma), human immunodeficiency virus, connective tissue disorders (e.g., scleroderma, Marfan syndrome, and

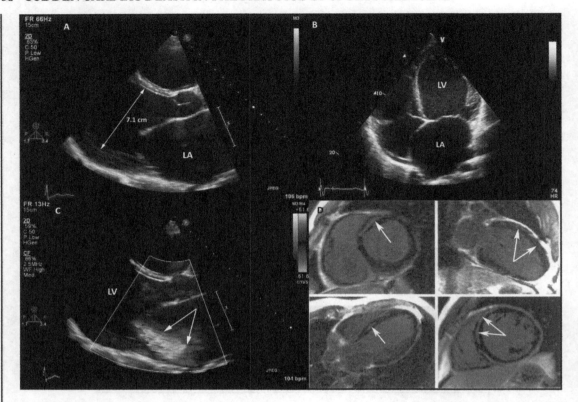

FIGURE 26: Echocardiographic features of DCM. A: Parasternal long-axis view showing gross LV dilatation of 7.1 cm; B: Apical 4-chamber view; note gross LA dilatation in addition to LV dilatation; C: Doppler image showing central jet of severe functional MR; D: CMRI images showing late gadolinium enhancement, taken from McCrohon et al. [254] with permission.

systemic lupus erythmatosus), and chronic alcohol abuse. Blood tests are performed to exclude the metabolic, inflammatory, and endocrine conditions, which associated with DCM. Once excluded, the patient is said to have idiopathic DCM. The ECG can be normal but may show non-specific abnormalities such as TWIs, septal Q-waves, axis deviation, or bundle branch block. The diagnosis is confirmed through standard 2D-echocardiography (Figure 26); diagnostic criteria include an ejection fraction of less 45%, fractional shortening of less than 25%, and left ventricular end-diastolic volume of greater than 112% of predicted [253]. Cardiac MRI may be useful in further quantifying LV enlargement and dysfunction, in addition to helping to determine the underlying etiology and detecting mid-wall fibrosis (Figure 26D) [254]. Conventional or computed tomography angiography is useful in excluding coronary artery disease as an etiological factor and for documenting the coronary artery anatomy. Patients with DCM are treated with standard heart failure medications including diuretics, angiotensin-converting enzyme inhibitors, beta-blockers, and aldosterone antagonists to both reduce symptoms and aid cardiac remodelling and long-term prognosis [158].

4.1.3.2 Dilated Cardiomyopathy in Sport. Given the fact that most individuals with DCM have a low functional capacity, it is not surprising that the prevalence of DCM is extremely low among athletes. Nevertheless, the differential diagnosis of DCM may still pose challenges for sports physicians in certain athletic cohorts. For example, it is well recognized that many endurance-trained athletes exhibit LV cavity dimensions exceeding 60 mm (and in some cases up to 70 mm), that are associated with a resting fractional shortening or ejection fraction at the lower limit of normal [255–257]. One study of 147 cyclists participating in the Tour de France found that 17 (11%) had a calculated LV ejection fraction <52% and impaired fractional shortening [255]. In addition, the ECG in DCM may be similar to that in athletic individuals, showing non-specific changes such as isolated voltage criteria for left ventricular hypertrophy. Ultimately, it is the ability of athletic

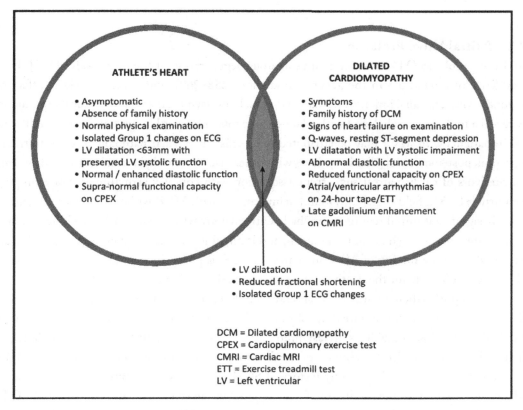

FIGURE 27: A practical guide for differentiating the physiological changes seen in athlete's heart from those of pathology in DCM.

individuals to augment cardiac output with exercise that is key in distinguishing between athlete's heart and DCM in such circumstances; the demonstration of a high peak oxygen consumption or normalisation of left ventricular systolic function during exercise are highly suggestive of physiological remodelling; however, other features can also aid this evaluation, and a diagnostic guide is shown in Figure 27.

Individuals with a definite diagnosis of DCM are disqualified from all competitive sports; however, those with a "low-risk" profile (namely, no symptoms, mildly depressed EF [>40%], normal BP response to exercise, no complex ventricular arrhythmias on 24-hour Holter monitoring) are eligible to participate in low-moderate dynamic and low-static sports [187] (see Table 4).

4.2   VALVULAR HEART DISEASE

Mitral valve prolapse and aortic stenosis account for up to 6% of all cases of SCD in young athletes.

4.2.1   Mitral Valve Prolapse

Mitral valve prolapse (MVP) is due to myxomatous degeneration of the mitral valve [188]. It has a prevalence of around 5% in the general population [258–260], and is more likely to affect tall individuals with a familial link [261]. Many individuals are asymptomatic, but physical examination may reveal a mid-systolic click and with a late systolic murmur, heard loudest at the apex. Although SCD in patients with MVP may be attributed to co-existent valvular pathology such as mitral regurgitation, post-mortem studies in patients with otherwise structurally normal hearts and evidence from survivors of cardiac arrest suggest an association with complex arrhythmias. For example, a review article by Kligfield et al. [258] found many patients with MVP, without evidence of hemodynamically significant mitral regurgitation, had ventricular arrhythmias on 24-hour Holter monitor. However, the precise mechanism is unknown, and it remains unclear whether a causal relationship exists or whether the finding of MVP in victims of SCD is purely incidental, with other causes of SCD being responsible for the death (such as ion channel disorders).

Work-up of patients with MVP should include echocardiogram to characterize the valve and assess for regurgitation (Figure 28), a 24-hour Holter monitor, and exercise stress test to assess for arrhythmias. Most individuals are asymptomatic, and current guidelines do not preclude participation in sport, provided the absence of the following: 1. symptoms (such as chest pain, palpitation or unheralded syncope); 2. family history of SCD; 3. documented ventricular arrhythmias; 4. moderate-to-severe mitral regurgitation; 5. associated Marfan syndrome; and 6. associated LQTS. Annual follow-up with echocardiography is recommended [188].

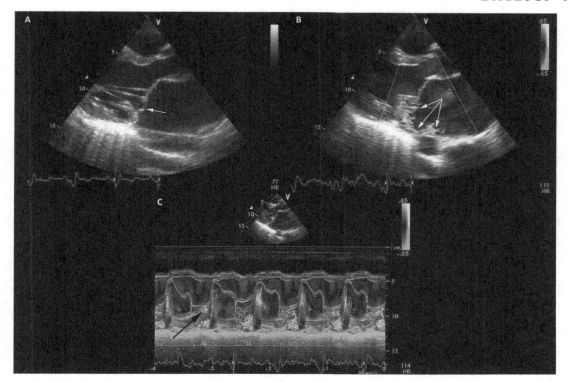

FIGURE 28: An echocardiographic study from a patient with MVP. A: Note prolapse of the anterior (A) mitral valve leaflet (arrow), along with thickening of the posterior leaflet (P); B: Doppler image showing associated posteriorly directed jet of severe mitral regurgitation; C: M-mode image through the valve, showing classic late systolic mitral regurgitation (arrow).

4.2.2 Aortic Stenosis

This is a rare but well-recognized cause of sudden death in athletes and one of the most common congenital cardiac lesions in adults. In young individuals, aortic stenosis is almost always associated with a congenital bicuspid aortic valve (BAV; Figure 29). Some individuals may develop aortic regurgitation secondary to a BAV rather than stenosis. Bicuspid aortic valves have various associations, with 50% of those affected having non-valvular lesions [262] including dilatation of the aortic root or thoracic aorta. It is estimated that 50%–75% of patients with BAVs have co-existent coarctation of the aorta [263]. A familial link has been found, with 9%–21% of probands having a first-degree relative also affected [264–267]. Furthermore, there appears to be a strong association with aortic dilatation in first-degree relatives, with one study finding over 30% with this abnormality

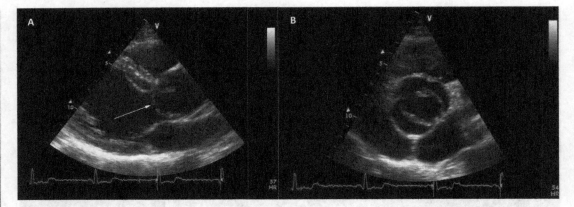

FIGURE 29: Echocardiographic features of a BAV. A: Eccentric closure line; B: Short-axis view, demonstrating 2 leaflets.

[268]. Within the past decade, developments have been made regarding the underlying genetics of BAVs. Garg et al. [269] and Mohamed et al. [270] both found a link with the NOTCH1 gene.

Symptoms of aortic stenosis include angina, breathlessness, and syncope. In those with moderate to severe stenosis or regurgitation, the condition will usually preclude an elite level of sporting performance. Athletes can be detected at screening through a history and physical examination, which may reveal an ejection systolic murmur and/or diastolic murmur. However, a large majority of young patients without valvular dysfunction will be asymptomatic and have no examination findings. Echocardiogram delineates the aortic valve extremely well and is the investigation of choice for establishing the diagnosis, assessing function, and long-term surveillance. Coexisting lesions such as root dilatation and coarctation of the aorta can also be discovered at echocardiography. The current recommendations regarding aortic valve lesions form the 36th Bethesda conference are shown in Table 8 [187]. Of note, patients with bicuspid aortic valves with no aortic root dilatation (less than 40 mm or the equivalent according to body surface area in children and adolescents) and no significant AS or AR may participate in all competitive sports [187].

4.3   DISORDERS OF THE AORTA AND CORONARY ARTERIES

According to some US data, congenital coronary artery anomalies (CCAA) account for up to one third of all cases of SCD and therefore represent an important cause [18, 34]. Premature coronary artery disease and aortic rupture are also responsible for a significant number of deaths; in some parts of Europe, the former is reported as the leading cause of death during sports in young adults [9].

TABLE 8: Eligibility criteria for athletes with aortic valve lesions. Taken from Maron et al. [187] with permission. NB: the reader is referred to Table 4 for the classification of sports.

LESION	RECOMMENDATIONS
Aortic Stenosis	1. Athletes with mild AS can participate in all competitive sports but should undergo serial evaluations of AS severity on at least an annual basis. 2. Athletes with moderate AS can engage in low-intensity competitive sports (class IA). Selected athletes may participate in low and moderate static or low and moderate dynamic competitive sports (classes IA, IB, and IIA) if exercise tolerance testing to at least the level of activity achieved in competition demonstrates satisfactory exercise capacity without symptoms, ST-segment depression, or ventricular tachyarrhythmias, and with a normal blood pressure response. Those athletes with supraventricular tachycardia or multiple or complex ventricular tachyarrhythmias at rest or with exercise can participate only in low-intensity competitive sports (class IA). 3. Patients with severe AS or symptomatic patients with moderate AS should not engage in any competitive sports.
Aortic Regurgitation	1. Athletes with mild or moderate AR, but with LV end-diastolic size that is normal or only mildly increased, consistent with that which may result solely from athletic training (12), can participate in all competitive sports. In selected instances, athletes with AR and moderate LV enlargement (60–65 mm) can engage in low and moderate static and low, moderate, and high dynamic competitive sports (classes IA, IB, 1C, IIA, IIB, and IIC) if exercise tolerance testing to at least the level of activity achieved in competition demonstrates no symptoms or ventricular arrhythmias. Those with asymptomatic non-sustained ventricular tachycardia at rest or with exertion should participate in low-intensity competitive sports only (class IA). 2. Athletes with severe AR and LV diastolic diameter greater than 65 mm as well as those with mild or moderate AR and symptoms (regardless of LV dimension) should not participate in any competitive sports.

TABLE 8: (continued)	
LESION	**RECOMMENDATIONS**
	3. Those with AR and significant dilation of the proximal ascending aorta (greater than 45 mm) can engage only in low-intensity competitive sports (class IA). These criteria do not apply to athletes with Marfan syndrome and AR, in whom the risks of aortic dissection and rupture are high, and any degree of aortic dilatation would be sufficient to prohibit competitive athletics.
Bicuspid Aortic Valve	1. Patients with bicuspid aortic valves with no aortic root dilatation (less than 40 mm or the equivalent according to body surface area in children and adolescents) and no significant AS or AR may participate in all competitive sports.
	2. Patients with bicuspid aortic valves and dilated aortic roots between 40 and 45 mm may participate in low and moderate static or low and moderate dynamic competitive sports (classes IA, IB, IIA, and IIB) but should avoid any sports in these categories that involve the potential for bodily collision or trauma.
	3. Patients with bicuspid aortic valves and dilated aortic roots greater than 45 mm can participate in only low-intensity competitive sports (class IA).

4.3.1 Congenital Coronary Artery Anomalies

Congenital coronary artery anomalies are present in up to 1% of the population, and the majority of individuals are largely asymptomatic, with many cases being found incidentally at angiography [271]. Several configurations are recognized, as illustrated in Figures 30 and 31, some of which carry high-risk characteristics [272]. Deaths from CCAA invariably occur during intensive physical exercise, which explains its relatively common incidence among athletes.

A large cohort study by Eckart et al. [273], investigating sudden death in American Army recruits who had been undergoing rigorous military training, found that one third of the deaths attributable to a structural heart condition at post-mortem were due to anomalous coronary arteries. In all of these cases, the abnormality discovered was the origin of the left coronary artery from the right sinus of Valsalva (Figure 30A; Figure 31C–F). In their retrospective multicenter review of

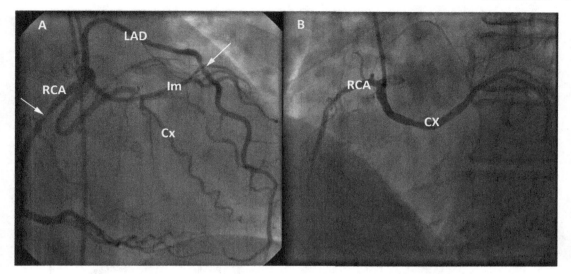

FIGURE 30: Examples of CCAAs. A: Entire left coronary system arising from the right sinus of Valsalva. Note the presence of co-existing coronary artery disease (arrows) in the right coronary artery (RCA) and left anterior descending (LAD). Im = intermediate branch. B: Example of an aberrant circumflex artery (Cx), arising from the right sinus of Valsalva.

deaths in athletes due to CCAA, Basso et al. [274] reported similar findings; 23 of the 27 sudden deaths due to anomalous coronary arteries arising from the wrong sinus of Valsalva were due to the left main coronary artery (LMCA) arising from the right aortic sinus, whereas only 4 were due to the right coronary artery (RCA) arising from the left aortic sinus (Figure 31A, B, and G). These findings may suggest a more malignant outcome in those patients with the left coronary artery arising from the right aortic sinus. Additionally, Basso et al. [274] found that one third of individuals with an anomalous LMCA origin had prodromal symptoms, whereas none of those with an anomalous origin of the RCA had complained of symptoms prior to sudden death.

The exact mechanism of sudden death is still open to debate, but various theories have been proposed. Most agree myocardial ischemia leads to a propensity to ventricular arrhythmias, but the exact mechanism is not fully understood. It has been suggested that this ischemic burden may not just be an acute insult at the time of sudden death but may be due to chronic myocardial ischemia, which results in myocardial necrosis and fibrosis [274]. This may leave a volatile electrical substrate, which under acute physical stress, can precipitate life-threatening arrhythmias. Several hypotheses have been suggested to explain the mechanism of the underlying ischemia. First, it may be due to the sharp angle that the anomalous coronary artery takes when arising from the aorta. Another

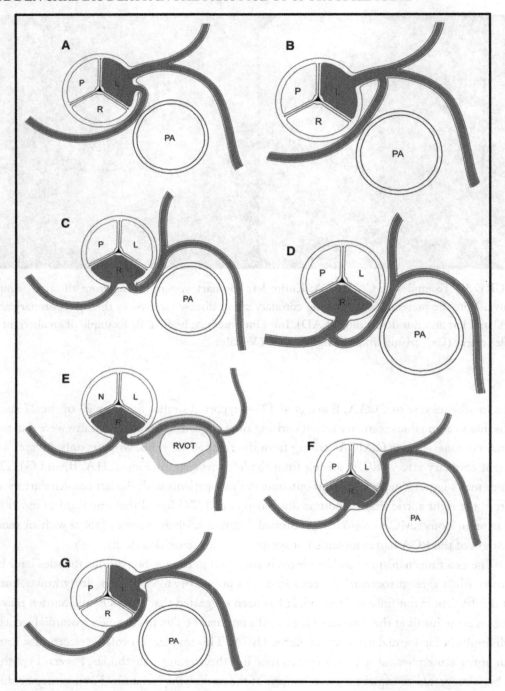

FIGURE 31: High-risk anomalous coronary artery patterns. PA = Pulmonary artery; P = posterior "noncoronary" sinus of Valsalva; L = left sinus of Valsalva; R = right sinus of Valsalva. A: Anomalous

theory suggests that it is due to compression of the anomalous coronary artery during exercise as it traverses between the aorta and the pulmonary trunk; indeed, several "high-risk" patterns have been suggested (Figure 31) [272]. A third explanation is that the anomalous coronary artery is predisposed to spasm due to endothelial injury [275].

Although two thirds of patients with CCAA are asymptomatic, in those with symptoms the most common are angina, syncope, and breathlessness. Unfortunately, identification of individuals with CCAA through screening programmes has proven problematic. Standard non-invasive investigations such as ECG and exercise stress testing are unlikely to pick up coronary artery anomalies, and therefore, their use is limited in the diagnosis of this condition, particularly in asymptomatic athletes [274]. Conventional coronary angiography [276], computed tomography [277], or CMRI [278] are much more sensitive and are advocated in those with symptoms in whom the diagnosis is being considered. Establishing the diagnosis and precise anatomic course of the anomalous vessel is particularly important in athletes given that current guidelines recommend disqualification from all competitive sports in those in whom the artery passes between the great arteries (due to the increased risk of SCD with this configuration) [279]. Surgical repair is indicated in all patients with coronary insufficiency and in asymptomatic patients with high-risk morphologic abnormalities (Figure 31) [271]. Repair options are dependent on the precise morphology and anatomy of the anomalous artery but include reimplantation, unroofing, and translocation (Figure 32).

According to the most recent guidelines, participation in all sports 3 months after successful operation is permitted for an athlete provided that there is no evidence of ischemia, ventricular or tachyarrhythmias, or dysfunction during maximal exercise testing [279].

---

right coronary artery (RCA) with separate ostium, arising from left coronary sinus (LCS), coursing interarterially, between the PA and aorta, without intramural course; B: Anomalous RCA with single ostium (shared with left coronary artery [LCA]), arising from LCS, coursing interarterially between the PA and aorta, without intramural course; C: Anomalous LCA with separate ostium, arising from right coronary sinus (RCS), coursing interarterially between the PA and aorta, without intramural course; D: Anomalous LCA with single ostium (shared with RCA), arising from right coronary sinus (RCS), coursing interarterially between the PA and aorta, without intramural course; E: Anomalous LCA with single ostium (shared with RCA), arising from the RCS, coursing interarterially and intramuscularly between the PA and aorta, without intramural course (RVOT = right ventricular outflow tract); F: Anomalous LCA with separate ostium, arising from the RCS, coursing interarterially and between the PA and aorta, with intramural course; G: Anomalous RCA with separate ostium, arising from the LCS, coursing interarterially between the PA aorta, with intramural course. Although not illustrated here, anomalous RCA with common ostium (shared with LCA) arising from the LCS, coursing interarterially between the PA and aorta, with intramural course, may occur. Modified from Gulati et al. [272] with permission.

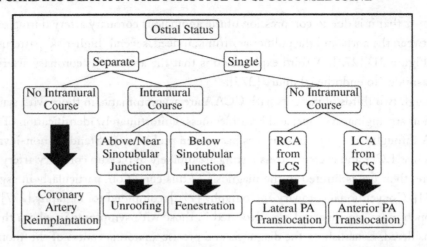

FIGURE 32: Examples of CCAAs. Morphology-based surgical management protocol. PA, Pulmonary artery; RCA, right coronary artery; RCS, right coronary sinus; LCA, left coronary artery; LCS, left coronary sinus. Taken from Gulati et al. [272] with permission.

4.3.2   Marfan Syndrome

Marfan syndrome is a multisystem inherited connective tissue disease with an incidence of 1 in 5000 live births. As those with Marfan syndrome are often extremely tall, they have an advantage in certain sports, in particular basketball and volleyball, and therefore the prevalence of the condition in these sports has been estimated to be much higher than in the general population [280]. Marfan syndrome affects all races and ethnic groups, and there appears to be no gender predilection. Although described initially just over a century ago by a French pediatrician Antoine Marfan, the genetic basis of the condition has only been understood over the past few decades. It is inherited in an autosomal dominant fashion and is due to a mutation in the fibrillin-1 gene located on chromosome 15. Fibrillin-1 is an extracellular matrix glycoprotein which is important for the maintenance of elastin fibers. Normal functioning of this protein is essential to maintain the structural integrity of connective tissue [281]. Although present throughout the body, elastin fibers are particularly abundant in the ligaments, parts of the eye and the aorta, which explains many of the clinical features seen in this condition. The extracellular matrix of connective tissue acts as a reservoir for growth factors including transforming growth factor beta (TGF-β) [281, 282]. Evidence from murine models suggest that abnormalities in fibrillin-1 lead to altered signalling of the TGF-β pathway and hence microfibril damage due to reduced sequestration of TGF-β [283, 284]. Further evidence that abnormal TGF-β signalling is responsible for the marfanoid phenotype has come from work in pediatric Marfan syndrome patients. A cohort study by Brooke et al. [285] followed up pediatric patients who were treated with angiotensin receptor blockers in combination with the standard therapy of beta-blockers. Although the numbers were small in this study (n=18), there appeared to be a reduction in aortic dilatation at follow-up. As angiotensin receptor blockers are known to inhibit TGF-β signalling, it is felt this was the likely mechanism of action.

Given the multisystem nature of the condition, patients can present in various ways. However, most patients do have a marfanoid phenotype, being very tall with long, thin limbs and an arm-span-to-height ratio of ≥1.05. They may also exhibit other stigmata including high arched palate, hyperextensibility, scoliosis, ectopia lentis, and skin striae (Figure 33). Patients may present to different specialists before the diagnosis is made. For example, they could present to a cardiologist with a murmur or cardiovascular symptoms or as an emergency with an acute complication such as aortic dissection or spontaneous pneumothorax to the emergency physician. Given the initial findings, an astute physician may retrospectively think of Marfan syndrome as an underlying cause. Alternatively, some patients have a family history, and therefore the diagnosis is made through screening.

Strict criteria have been developed in order to make the diagnosis of Marfan syndrome. The most common, internationally used diagnostic classification system is the Ghent criteria [280], which were established by a group of experts in 1995 and recently updated in 2010 (Table 9) [286]. These criteria include various major and minor features, which a suspected individual needs to fulfil in order to warrant a diagnosis of Marfan syndrome.

Cardiovascular complications occur in a high proportion of those with Marfan syndrome. It is estimated that by the age of 21 years, at least 50% of patients will have evidence of cardiovascular sequelae [287] and that 70% of Americans with the condition will die from these complications [288]. Cystic medial necrosis within the connective tissue of the vasculature weakens the vascular wall predisposing to aortic dilatation, aneurysms, aortic dissections, and rarely aortic ruptures. This is compounded by stiffness and poor compliance of the ascending, descending, and abdominal aorta.

Individuals known to have Marfan syndrome should undergo regular surveillance for evidence of aortic root dilatation and aneurysm formation, often having prophylactic aortic root replacement when dilatation becomes significant (>45–50 mm; Figure 34). Any patient suspected of having Marfan syndrome should have an echocardiogram to assess the aortic root and heart valves. The patient should also have a genetic referral, which should include genetic testing for the diagnostic mutation in the fibrillin-1 gene.

Occasionally, individuals who are not known to have Marfan syndrome may present with an aortic dissection as a surgical emergency or may present with sudden death secondary to aortic rupture. Valvular heart disease is also a significant cause of morbidity and mortality in these patients. Those with progressive aortic root dilatation are predisposed to aortic regurgitation, which may

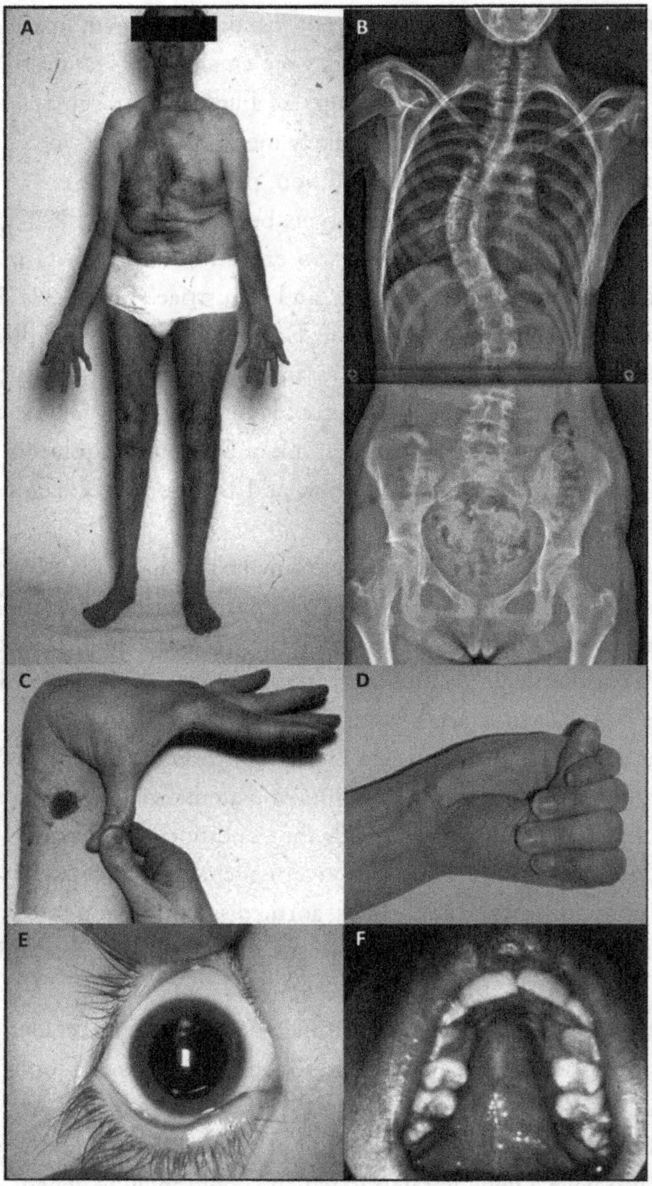

FIGURE 33: Clinical manifestations of Marfan syndrome. A: Typical Marfan body habitus. Note tall stature and long thin extremities; B: X-ray showing marked scoliosis; C. Hyperextensibility of the joints; D: Positive Steinberg thumb sign (thumb extends beyond lateral border of palm); E: Ectopia lentis; F: High-arched palate with crowding of teeth. With kind permission from Dr. Anne Child, St. George's University of London.

TABLE 9: Revised Ghent criteria for diagnosis of Marfan syndrome and related conditions. Taken from Loeys et al. [286] with permission.

In the absence of family history:
(1) Ao (Z ≥ 2) AND EL = MFS*
(2) Ao (Z ≥ 2) AND *FBN1* = MFS
(3) Ao (Z ≥ 2) AND Syst (≥7pts) = MFS*
(4) EL AND*FBN1* with known Ao = MFS

EL with or without Syst AND with an *FBN1* not know with Ao or no *FBN1* = ELS
Ao (Z < 2) AND Syst (≥5 with at least one skeletal feature) without EL = MASS
MVP AND Ao(Z < 2) AND Syst (<5) without EL-MVPS

In the presence of family history:
(5) EL AND FH of MFS (as defined above) = MFS
(6) Syst (≥ 7 pts) AND FH of MFS (as defined above) = MFS*
(7) Ao (Z ≥ 2 above 20 years old, ≥3 below 20 years) + FH of MFS (as defined above) = MFS*

*Caveat: without discriminating features of SGS, LDS, or vEDS (as defined in table 1) AND after *TGFBR1/2*, collagen biochemistry, *COL3A1* testing if indicated. Other conditions/genes will emerge with time.

Ao, aortic diameter at the sinuses of Valsalva above indicated Z-score or aortic root dissection; EL, ectopia lentis; ELS ectopia lentis syndrome; *FBN1*, fibrilin-1 mutation (as defined in box 3); *FBN1* not known with Ao, *FBN1* mutation that has not previously been associated aortic root aneurysm/dissection; *FBN1* with known Ao, *FBN1* mutation that has been identified in an individual with aortic aneurysm; MASS, myopia, mitral valve prolapse, borderline (Z < 2) aortic root dilatation, striae, skeletal findings phenotype; MFS, Marfan syndrome; MVPS mitral valve prolapse syndrome; Syst, systemic score (see box 2); and Z, Z-score.

necessitate aortic valve replacement often with simultaneous aortic root replacement [280]. In the pediatric and adolescent Marfan population, the most common valvular abnormality found is mitral valve prolapse. Although this may be relatively benign in some, in others, it may lead to severe mitral regurgitation, which may in turn require mitral valve replacement or repair [289].

When the diagnosis of Marfan syndrome is made in childhood, parents are advised to discourage their child from playing high-impact, high-intensity sports (see Table 4) and rather

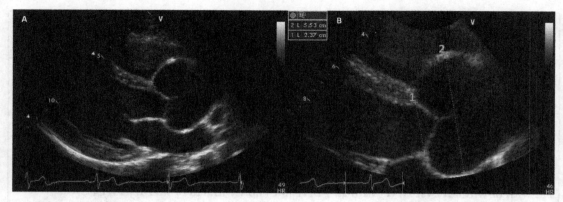

FIGURE 34: Echocardiogram from a patient with Marfan syndrome. A: Parasternal long axis view showing dilated, bulbous aortic root; B: The diameter measured 55.3 mm, and the patient went on to have a root replacement.

encourage them to participate in more appropriate activities such as sailing or archery. If an individual is gene positive or fulfils the Ghent criteria, they should avoid playing competitive sports involving prolonged exertion at peak capacity. Those with predominantly skeletal abnormalities can participate in low-intensity, leisure activities but need to avoid contact sports due to the risk of damage to the aorta or eyes. In the absence of aortic dilation, those with MVP can also participate in non-contact, moderate intensity sports such as running, swimming, or playing tennis [188, 279].

4.3.3   Premature Coronary Artery Disease

Although coronary artery disease is the most common cause of SCD in those over 35 years, it is an uncommon cause of sudden death in young patients and athletes. Most cases are likely due to familial hypercholesterolemia and have a prevalence in the general population of around 0.2% [41]. Nearly 4% of sudden deaths in athletes can be attributed to premature coronary artery disease [12, 31, 39], although in certain countries such as Norway the incidence appears to be higher [9]. Many patients are asymptomatic but some may have symptoms of angina. Some individuals are identified through screening as a result of a family history of premature coronary artery disease or sudden death. Clinical examination is likely to be normal, but rarely, some of the peripheral stigmata of a familial hyperlipidemia syndrome may be present such as xanthelasma, corneal arcus, palmer, and eruptive xanthomata. Treatment involves aggressive management of risk factors, most importantly hyperlipidemia, with lipid-lowering agents. The confirmation of ischemic heart disease with a high probability of cardiac events disqualifies the individual from any competitive sports. Those with ischemic heart disease but a low probability of cardiac events are allowed to continue low-moderate dynamic and low-static sports [188].

4.4 PRIMARY ELECTRICAL DISORDERS OF THE HEART

Sudden cardiac death can occur in patients with structurally normal hearts. These cases are usually due to underlying electrical disorders such as to ion channel diseases or accessory pathways. These will now be explored further.

4.4.1 Long-QT Syndrome

The long-QT syndromes (LQTSs) are characterized by prolonged ventricular repolarization, predisposing to polymorphic VT and sudden death. Thirteen different forms of congenital LQTSs are recognized, named LQTS 1–13. Two major variants exist as follows: the autosomal dominant Romano–Ward syndrome and the rare autosomal recessive Jervell and Lange–Nielsen syndrome. Each type of LQTS is caused by mutations at specific gene loci, which encode for various sodium (Na^+), potassium (K^+), and calcium (Ca^{2+}) channels. Congenital LQTS has an overall prevalence of approximately 1 in 2500 [290], but affected family members may show a borderline or even normal QTc, and hence, this incidence may be underestimated. Of those patients, 5%–10% present with aborted SCA or SCD.

Long-QT syndrome 1 is due to a loss of function mutation in KCNQ1, the gene encoding for $K_v7.1$, a protein comprising the α subunit of the delayed rectifier K^+ channel (see Figure 35 for an example of a generic K^+ channel) [291]. This leads to a reduction in I_{Ks} which is responsible for the prolonged action potential duration and prolongation of the QT interval [292]. The majority of cardiac events appear to be precipitated by physical exertion, with 1 study noting that one third of those with an obvious precipitant had been swimming when the event occurred [293]. Long-QT syndrome 2 is caused by a mutation in the KCNH2 gene (also known as the HERG gene). This encodes for the α-subunit ($K_v11.1$) of the K^+ channel, which is responsible for the rapidly activating delayed outward rectifier current I_{Kr}. Emotional stress, including loud auditory stimuli, appears to precipitate events in patients with this subtype [293]. Long-QT syndrome 3 is caused by mutations in the gene coding for the α-subunit of the cardiac fast Na^+ channel SCN5A (Figure 25). It is felt that SCN5A mutations delay repolarization by enhancing the small late Na^+ (I_{NaL}) current, which occurs in phases 2 and 3 of the cardiac action potential. This enhanced I_{NaL} may trigger after depolarizations due to the reactivation of the L-type Ca^{2+} channel (Figure 26). Unlike LQT1 and LQT2, cardiac events seem to occur at rest or during sleep [293]. It is thought that 90% of the genotypes of LQTS subjects have mutations causing LQT1, LQT2, or LQT3 [291].

Long-QT syndrome 4 is a rare entity and unlike the other forms of LQTS is not caused by loss of function of an ion channel gene but instead is due to a mutation in the gene encoding

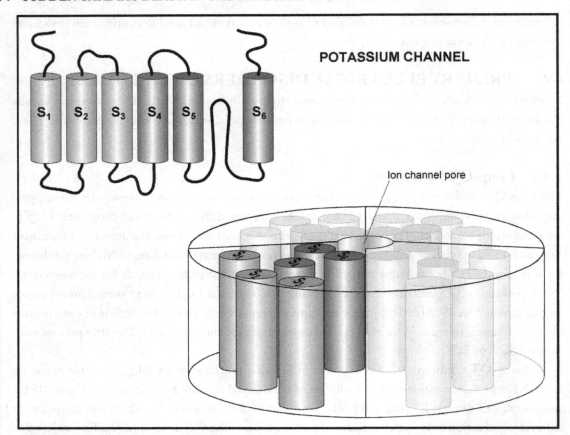

FIGURE 35: Schematic representation of a generic cardiac potassium channel.

for ankyrin-B. This leads to disorganization within the cell of the Na^+/Ca^{2+} exchanger, Na^+/K^+ ATPase, and the inositol 1-4-5 triphosphate (IP3) receptors. The resulting increased intracellular Na^+ leads to after-depolarisations, predisposing to arrhythmias when the individuals is exposed to catecholamine stress [294]. Other uncommon LQTSs include LQT5 and LQT6, which are both caused by abnormal functioning of K^+ channels. Long-QT syndrome 5 is due to a loss of function mutation in the KCNE1 gene, which encodes for the regulatory β-subunit (Mink) of the delayed rectifier K^+ channel. This normally functions to regulate the inward low K^+ current (IKs). Although LQT5 is much rarer, these patients are phenotypically similar to those with LQT1 [292]. Loss of function mutations of the KCNE2 gene, encoding for MiRP1, reduces the rapidly activating delayed outward rectifying current and is responsible for LQT6 [292]. Long-QT syndrome 7, also known as Andersen–Tawil syndrome, is a condition characterized by periodic paralysis, marked

QT-prolongation, and dysmorphic features. It is due to a mutation of the gene responsible for the inward rectifying K^+, KCNJ2, which is found at the 17q23 locus [295]. Long-QT syndrome 8, or Timothy syndrome, is caused by dysfunction of the $Ca_v1.2$ channel (Figure 36) by a missense mutation G406R in the vast majority of cases. In cardiac myocytes, this leads to a continuous inward Ca^{2+} current due to loss of voltage-dependent channel inactivation. This G406R missense mutation is expressed in all affected tissues and is likely to be responsible for the multisystem nature of this condition. Those affected display distinctive characteristics such as marked QT-prolongation, syndactyly, congenital heart defects and developmental disorders [297]. The other LQTSs (9–13) are very rare and outlined in Table 10.

In the main, mutations in LQTS are missense mutations; however, they do include small intragenic deletions, splice errors, nonsense mutations, insertions, and deletions. Although over 700 mutations have been discovered [142], one study by Priori et al. [297] suggested that more than half of these are caused by 64 mutations of the KCNQ1, KCNH2, and SCN5A genes. Furthermore, this study suggested using these mutations as a starting point when screening for genetic abnormalities

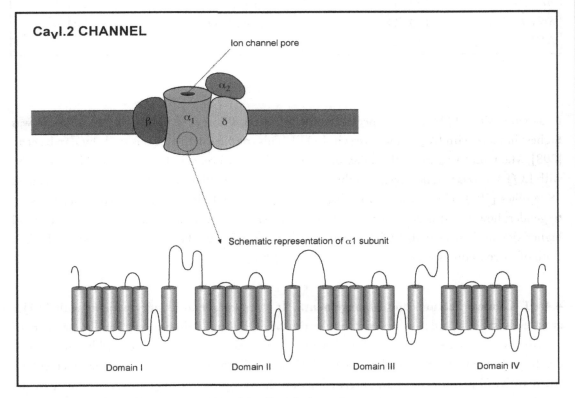

FIGURE 36: Schematic representation of the $Ca_v1.2$ channel.

TABLE 10: Genetic basis and frequency of LQTSs.
Taken from Barsheshet et al. [299] with permission.

GENE	LQTS SUBTYPE	PROTEIN	FREQUENCY (%)
KCNQ1	LQT1	Kv7.1	30–35
KCNH2	LQT2	Kv11.1	25–30
SCN5A	LQT3	NaV1.5	5–10
ANKB	LQT4	Ankyrin	<1
KCNE1	LQT5	MinK	<1
KCNE2	LQT6	MiRP1	<1
KCNJ2	LQT7	Kir2.1	<1
CACNAIC	LQT8	L-type calcium channel	<1
CAV3	LQT9	Caveolin 3	<1
SCN4B	LQT10	Sodium channel β4-subunit	<1
AKAP9	LQT11	Yotiao	<1
SNTA1	LQT12	Syntrophin-α1	<1
KCNJ5	LQT13	Kir3.4	<1

in patients with LQTS or in suspected family members. Priori et al. found the risk of SCD was highest in those with LQT2 and lowest in LQT1. This conflicts with previous study by Zareba et al. [298], who found a significantly higher proportion of cardiac events in LQT1 and LQT2 compared with LQT3. It was felt, however, that this study may have potentially been biased as it only included 38 families [297]. The risk of first cardiac arrest/sudden death does appear to have a relationship to gender; however, it is dependent on the specific genetic locus described. The same study found higher risk in females with LQT2 compared to males, whereas the converse was true in LQT3. Both of these findings were statistically significant [297].

4.4.1.1 Clinical Features and Management. It is important to identify patients with LQTS at an early stage in order to ensure that the risk of SCD can be minimized. Family members of a relative who has died of the condition or first-degree relatives of a victim of sudden arrhythmic death syndrome (SADS) should be screened. Occasionally, patients may be identified incidentally when an ECG performed for an unrelated reason shows a prolonged QT interval. Some patients may have symptoms such as palpitations, dizziness, or unheralded syncope. A careful history should be undertaken, ascertaining any family history of sudden or unexplained death. This may include events such as drowning, other unexplained accidents, and epileptiform seizures.

In the normal resting ECG, the upper limit of normal for the QT interval corrected for heart rate (corrected QT, QTc) is 440 msec in males and 460 msec in females [300]. Aside from prolongation of the QT interval, abnormal T-wave morphologies may also be seen in the LQTSs (Figure 37; Table 11) [301]. A useful tool in determining a diagnosis of LQTS is the modified Schwartz score (Table 11), which divides the probability of LQTS into low, intermediate, and high based on a combination of clinical and electrocardiographic features [301]. Other investigations can help to identify the broader phenotypic features of LQTS. For example, 24-hour Holter monitoring may identify paroxysms of NSVT (particularly of *torsade de pointes* morphology), and exercise stress testing may reveal paradoxical prolongation of the QT interval at higher heart rates [302].

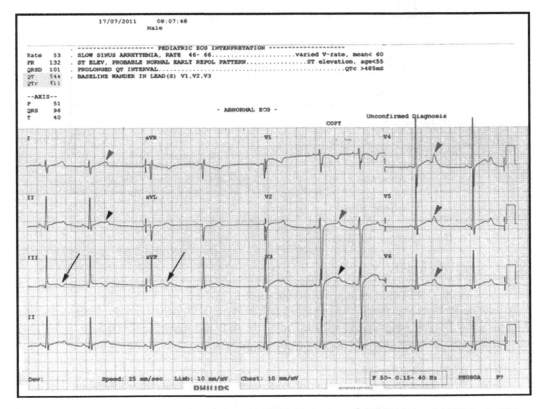

FIGURE 37: An ECG from a patient with LQTS. Note a corrected QT interval (QTc) of >500 msec (highlighted), along with morphological T-wave changes including bi-phasic T-waves (black arrows), late-peaking T-waves (red arrowheads), and notched T-waves (black arrowheads).

TABLE 11: Modified Schwartz criteria for the diagnosis of LQTS. Taken from Schwartz et al. [301] with permission.

		POINTS
ECG findings*		
A.	$QT_c^\dagger$	
	≥480 msec$^{1/2}$	3
	460–470 msec$^{1/2}$	2
	450 msec$^{1/2}$ (in males)	1
B.	Torsade de pointes‡	2
C.	T-Wave alternans	1
D.	Notched T wave in three leads	1
E.	Low heart rate for age§	0.5
Clinical history		
A.	Syncope‡	
	With stress	2
	Without stress	1
B.	Congenital deafness	0.5
Family history$^\Vert$		
A.	Family members with definite LQTS$^\#$	1
B.	Unexplained sudden cardiac death below age 30 among immediate family members	0.5

LQTS, long QT syndrome

*In the absence of medications or disorders known to affect these electrocardiographic features.

†QT$_c$ calculated by Bazett's formula, where QT$_c$ = QT/$\sqrt{RR}$.

‡Mutually exclusive.

§Resting heart rate below the second percentile for age [25].

$^\Vert$The same family member cannot be counted in A and B.

$^\#$Definite LQTS is defined by an LQTS score ≥ 4.

Scoring: ≤1 point, low probability of LQTS; 2 to 3 points, intermediate probability of LQTS; ≥4 points, high probability of LQTS.

This investigation may be more useful in those with suspected LQT1 and LQT2, as sudden death in these cases is often related to exercise [303, 304]. Assessment of the QTc during exercise is most useful during a heart rate rise from baseline to 120 beats per minute; however, attention should be paid right into the recovery phase, as the QT interval may not prolong until this time. In one study, the best separation between patients with LQTS and controls was obtained at a cutoff point of QT prolongation of 41 msec 3 minutes into recovery [302].

Beta-blockers are the mainstay of therapy for all patients with LQTS regardless of the subtype, although they do appear to have much greater effect in those with LQT1 and LQT2 [304–306]. Patients are also given advice on lifestyle modification, which includes avoidance of sports with high adrenaline surges, in particular swimming (see section 4.4.1.2 below) [41]. The use of ICDs is controversial in LQTS; they are generally advised for secondary prevention in survivors of cardiac arrest, patients with sustained VT, and those who have syncope despite pharmacological intervention [307, 308]. Outside of these scenarios, the decision regarding ICD implantation is made on a case-to-case basis.

4.4.1.2 Long-QT Syndromes and Sport. In athletes, the diagnosis of LQTS can prove challenging given that accurate calculation of the QT interval may be hindered by bradycardia, sinus arrhythmia, and prominent U waves, all of which are frequently observed as part of the physiological response to exercise [82, 97]. In addition, athletes have been noted to have a longer QT interval than sedentary individuals, which again is likely to be part of cardiac adaptation to exercise [309]. Therefore, in order to accommodate these observations, current recommendations suggest an upper limit of normal of 470 msec in male athletes and 480 msec in female athletes [279]. A clear-cut diagnosis is most easily established in an athlete when the corrected QT interval is >500 msec [97, 309, 310]. In borderline cases (i.e., a QTc of 480–500 msec), the diagnosis can prove challenging. The physician therefore needs to rely on other supporting information to establish the diagnosis (Table 11) [41, 301]. In these borderline cases, genetic testing may prove particularly useful in an athlete, not only allowing the diagnosis to be established but also confirming a particular phenotype and helping to guide therapy [311]. Certain phenotypes are particularly relevant to the athlete; for example, in LQTS 1, physical exertion (particularly swimming) appears to be a common trigger for ventricular arrhythmias. In contrast, individuals with LQT2 appear more at risk from auditory/emotional triggers, whereas those with LQT3 appear to be at greater risk during rest and inactivity [279].

Current guidelines for participation in sport in those with LQTS state the following (taken from Maron et al. [279] with permission):

1. Regardless of QTc or underlying genotype, all competitive sports, except those in class IA category (see Table 4) should be restricted in a patient who has previously experienced

either (1) an out-of-hospital cardiac arrest or (2) a suspected LQTS-precipitated syncopal episode.

2. Asymptomatic patients with baseline QT prolongation (QTc of 470 msec or more in males, 480 msec or more in females) should be restricted to class IA sports. The restriction limiting participation to class IA activities may be liberalized for the asymptomatic patient with genetically proven type 3 LQTS (LQT3).

3. Patients with genotype-positive/phenotype-negative LQTS (i.e., identification of a LQTS-associated mutation in an asymptomatic individual with a nondiagnostic QTc) may be allowed to participate in competitive sports. Although the risk of sudden cardiac death is not zero in such individuals, there is no compelling data available to justify precluding these individuals (who are being identified with increasing frequency) from competitive activities. Because of the strong association between swimming and LQT1, persons with genotype-positive/phenotype-negative LQT1 should refrain from competitive swimming.

4. LQTS patients with an ICD/pacemaker should not engage in sports with a danger of bodily collision because such trauma may damage the pacemaker system. The presence of an ICD should restrict individuals to class IA activities.

We now present a case study highlighting the difficulties that may be encountered in clinical practice when assessing athletes with ECG features of LQTS.

4.4.1.3 Long-QT Syndrome Case Study. A 15-year-old female Caucasian professional football player was seen at a pre-participation screening event. Her ECG is shown in Figure 38; she was noted to have TWIs in leads V1–V3 and lead III (arrows) along with a slightly prolonged corrected QT interval (QTc; 467 msec) for her gender (upper limit of normal ≤460 msec in a female). In addition, late-peaking T-waves were seen in leads V4–V6 (arrowheads). She was entirely asymptomatic, with no medical history of palpitation, syncope, and no family history of cardiac disorders or sudden death. There was no personal or family history of seizures, drowning or near-drowning, or unexplained road traffic accidents. At this point, it was felt that the TWIs may represent a persistent juvenile pattern, and she was therefore advised to return a year later.

Her ECG 1 year later, aged 16, is shown in Figure 39 below; she had remained asymptomatic in the interim. Note the persistence and in fact extension of her TWIs in V1–V4 (arrows), along with late-peaking T-waves in V5–V6 (arrowheads). Her QTc remained at the upper limits of normal at around 459 msec. Given these findings, she was invited to a hospital for further investigations. At this point, the concern was of ARVC, although the differential diagnosis included that of LQTS. A repeat ECG was performed, which is shown in Figure 40 below.

Note the resolution of some of the TWIs; comprehensive evaluation including a signal-averaged ECG, echocardiogram, 24-hour Holter monitor, and CMRI scan showed no evidence of

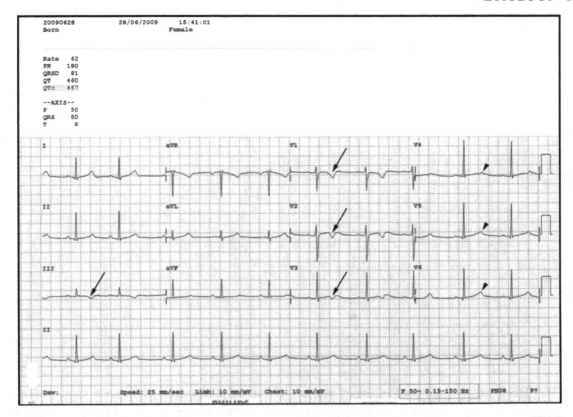

```
20090628          28/06/2009      15:41:01
Born                              Female

Rate    62
PR     180
QRSD    81
QT     460
QTc    467

--AXIS--
P       50
QRS     50
T        8
```

I   aVR   V1   V4
II   aVL   V2   V5
III aVF   V3   V6
II

```
Dev:          Speed: 25 mm/sec   Limb: 10 mm/mV   Chest: 10 mm/mV      F 50~ 0.15-150 Hz     PH08     P?
```

FIGURE 38: An ECG from a 15 year old female Caucasian professional football player. Note TWIs in leads V1–V3 and lead III (arrows) along with a slightly prolonged corrected QT interval (QTc; 467 msec). In addition, late-peaking T-waves were seen in leads V4–V6 (arrowheads).

ARVC. However, note the emergence of other ECG features, including a now prolonged QTc of 480 msec and late-peaking T-waves in leads V3–V5 (see arrowheads). Genetic testing was offered to the athlete but declined; however, she had a normal exercise stress test (no evidence of paradoxical prolongation of the QT interval) and 24-hour Holter monitor. In addition, ECGs on both her parents were normal, with no morphological T-wave changes and QT intervals well within the normal limits.

4.4.1.4 Discussion to Case Study.   This case illustrates the complexities that can exist when interpreting an athlete's ECG, which can often change over time, especially during adolescence. In particular, the value of obtaining serial ECGs is clearly illustrated. The differential diagnosis of TWIs is highlighted, which are not only features of a cardiomyopathy but can also be manifestations of other conditions including ion channel disorders.

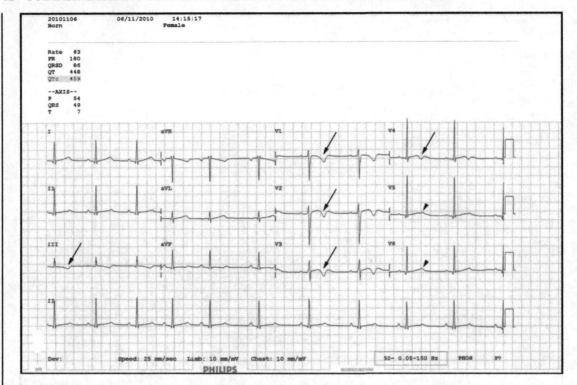

FIGURE 39: ECG from the footballer 1 year later. Note the persistence and in fact extension of her TWIs in V1–V4 (arrows), along with late-peaking T-waves in V5–V6 (arrowheads). Her QTc remained at the upper limits of normal at around 459 msec.

Athletic training not only results in TWIs but is also associated with prolongation of the QT interval [334]. It is possible that this athlete's TWIs represented a persistent juvenile pattern, evidenced by the fact that they were diminishing with age. However, it must be remembered that LQTS can be a complex diagnosis to make, with not only abnormalities in the QT interval itself but also complex changes in T-wave morphology (see Table 11 modified Schwartz criteria).

Usually, a concrete diagnosis of LQTS can only be established in those athletes with a QTc >500 msec [334]. In borderline cases such as these, genetic testing may prove extremely useful, although can be time consuming and therefore impractical. In the case, the athlete declined genetic testing for personal reasons; however, her QTc was <500 msec, she has no broader phenotypic features of LQTS, and her parents' ECGs were completely normal. She was therefore cleared to play professionally but with annual follow-up and made aware of symptoms that would warrant immediate re-assessment.

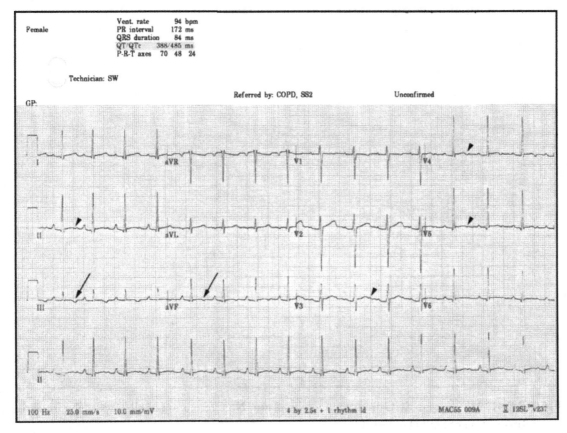

FIGURE 40: Repeat ECG aged 17.

The player returned a year later at age 17 years, at which point she had injured her ankle. A repeat ECG was performed which is shown in Figure 41. Note the total resolution of her ECG changes and T-wave abnormalities, illustrating further the transient changes that may occur during adolescence, particularly when combined with intense exercise.

4.4.2   Short-QT Syndrome

Short-QT syndrome (SQTS) is a condition which is largely in its infancy, being described by Gussak et al. [312] in 2000. The condition is characterized by a persistently short-QT interval (<300 msec) and tall, symmetrical peaked T-waves on the resting ECG (Figure 42). As SQTS is a relatively new clinical entity, not much is known about the natural history of the condition; however,

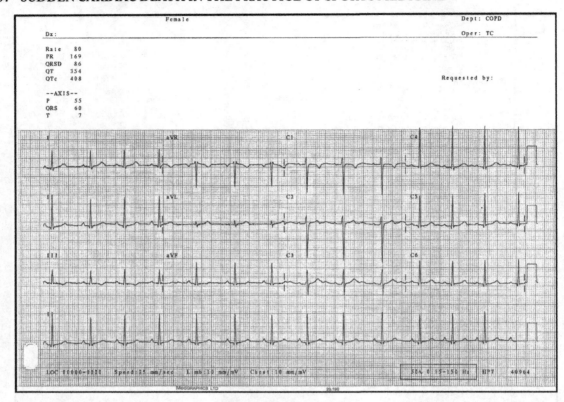

FIGURE 41: Most recent ECG from the football player. Note resolution of all T-wave abnormalities and normalization of her QT interval.

the clinical picture encompasses syncope, palpitations, malignant cardiac arrhythmias, and sudden death [191]. Diagnosis is made when a patient presents with the typical ECG findings in the presence of a structurally normal heart and inducible ventricular fibrillation on electrophysiological study [299]. So far, 20% of patients appear to have a positive genotype with mutations either being a gain of function mutation in the delayed K^+ rectifier channel or a loss of function mutation in the gene encoding for the L-type Ca^{2+} channel. Five different subtypes of SQTS are recognized. Short-QT syndrome 1 is the commonest form, caused by a gain of function mutation in KCNH2, which affects the K^+ channel, responsible for the rapidly activating delayed outward rectifier K^+ current. This results in an increased I_{Kr}, accelerating repolarization, shortening the action potential, and acting a substrate for arrhythmias. So far, 3 mutations have been identified in SQTS1 as follows: N558K, E50D, and R1135H. Short-QT syndrome 2 is due to a mutation in KCNQ1, which affects the I_{Ks}. Short-QT syndrome 3 is due to a mutation in KCNJ2, which affects the I_{K1} [299,

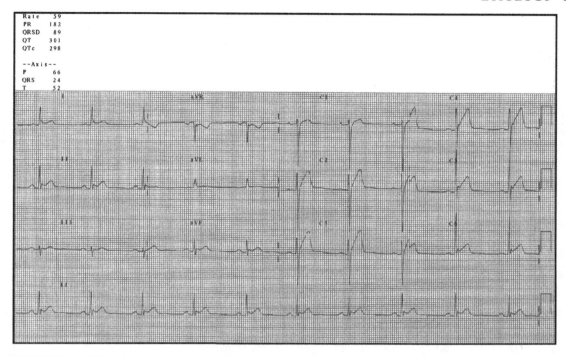

Rate   59
PR   182
QRSD   89
QT   301
QTc   298

--Axis--
P   66
QRS   24
T   52

FIGURE 42: ECG from a patient with SQTS. Note the QTc of 300 msec and typical tall, symmetrical, peaked T-waves.

313, 314]. Although SQTSs 1–3 are due to different mutations of the cardiac K^+ channels, they all ultimately result in a greater efflux of K^+ from the cardiomyocyte during the repolarization, shortening the QT interval and predisposing to arrhythmias. Short-QT syndrome 4 is due to a mutation in the gene encoding for the L–type Ca^{2+} channel (CACNA1C), whereas SQTS5 is due to a mutation of its β–subunit, encoded for by CACNB2b [315]. Like many of the inherited arrhythmic diseases, treatment is limited to risk stratification and ICD insertion on those deemed high risk (including those with aborted cardiac arrest) [291].

Given that the understanding SQTS is still in its infancy, current recommendations for athletes advocate universal restriction from all competitive sports, with the possible exception of class IA activities [279].

4.4.3   Brugada Syndrome

Brugada syndrome (BrS) is a rare cardiac condition first described by Brugada and Brugada in 1992 [316]. Classically, it is characterized by a coved ST-segment elevation in the right precordial leads

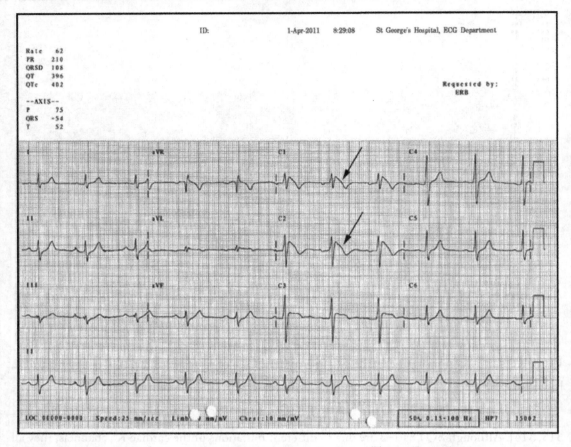

FIGURE 43: An ECG from a patient with BrS. Note the typical Type 1 Brugada phenotype in leads V1–V2, with "coved" ST-segment elevation (black arrows).

V_1–V_3 on the resting ECG ("Type 1" ECG pattern; Figure 43) [291], although in some patients this phenotype may be concealed and only induced by provocation testing. Affected individuals have a propensity to SCA secondary to polymorphic VT, which often occurs at rest or during sleep. Brugada syndrome has a prevalence of 5 in 10,000 and is inherited in an autosomal dominant pattern [291] but with a strong male predilection [317, 318]. Animal studies suggest the gender difference is due to a more pronounced transient outward K^+ current (I_{to}) in males [319]. Brugada syndrome is responsible for >4% of SCD cases and >20% in those with structurally normal heart [317]. Risk factors for cardiac events include male gender, a spontaneous Type 1 pattern on the resting ECG [320], unheralded syncope, and inducible VT during programmed ventricular stimulation [321]. High fevers appear to precipitate SCA in children and adults [322, 323]. Typically, BrS occurs in

TABLE 12: Genetic basis and frequency of BrS. Taken from Bastiaenen and Behr [327] with permission.

BRS SUBTYPE	GENE	PROTEIN	ION CHANNEL	EFFECT OF MUTATION	FREQUENCY
BrS1	SCN5A	$Na_v1.5$	αsubunit/ Na	Loss of function	21%
BrS2	GPD1L	G3PD1L	Interacts with	Loss of function	—
BrS3	CACNA1C	$Ca_{v1.2}$	α subunit/ Na	Loss of function	5.5%
BrS4	CACNB2	$Ca_v\beta2$	β subunit/ L, Ca	Loss of function	4.9%
BrS5	SCN1B	$Na_v\beta$	β subunit/ L, Ca	Loss of function	—
BrS6	KCNE3	MiRP2	β subunit/ Na	Loss of function	—
BrS7	SCN3B	$Na_v\beta3$	β subunit/ KS/to	Loss of function	—
BrS8	CACNA2D1	$Ca_v\alpha2\delta1$	α subunit/ Na $\alpha2\delta$ subunit/L, Ca	Loss of function	1.8%

structurally normal hearts [324], although a subpopulation of ARVC patients have been found to display an ST-segment elevation and polymorphic VT that is characteristic of Brugada syndrome [317, 325]. Of interest, recent studies have suggested that some SCN5A defects may be capable of causing fibrosis in the conduction system and ventricular myocardium [317, 326].

Currently, 8 subtypes of BrS are recognized (Table 12) [327]. Brugada syndrome 1 comprises between 18% and 30% of cases and is due to loss of function mutations of SCN5A [317]. The resulting electrophysiological abnormalities are caused by 2 major mechanisms as follows: reduced I_{Na} (entry of Na^+ channel into an intermediate state of inactivation, which recovers more slowly) and accelerated inactivation of the Na^+ channel. Brugada syndrome 2 is due to a mutation in GPD1L, which encodes for glycerol-1–Phosphate dehydrogenase 1-like protein (G3PD1L). Brugada syndrome 3 is due to mutations in CACNA1C, which encodes for the pore-forming $\alpha1$ subunit of the $Ca_v1.2$ channel (Figure 44).

The 2 most common mutations, G490R and A39V, are both missense mutations, which lead to an increase in the slow long Ca^{2+} current and therefore prolong repolarization [290]. Brugada syndrome 4 is due to a loss of function mutation in CACNB2, which encodes for $Ca_v\beta2$, the $\beta2$ subunit of the Cav1.2 ion channel (Figure 36). This CACNB2 mutation interferes with the stimulatory role of $Ca_v\beta2$ and as a result reduces the depolarising $I_{L, Ca}$. It has been proposed that mutations lead to increased transmural dispersion of repolarization across the ventricular wall, which leads to a predisposition to a re-entry phenomenon [291]. Mutations in the Ca^{2+} channels

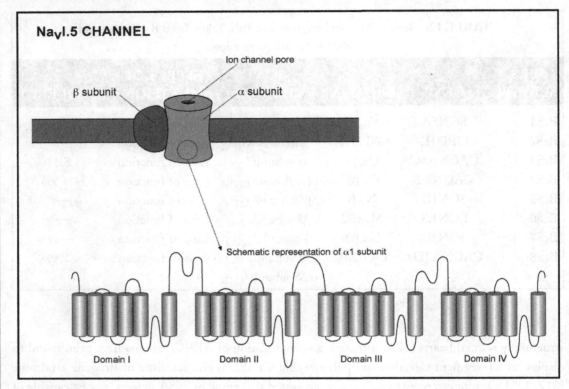

FIGURE 44: Schematic representation of the $Na_v1.5$ channel.

CACNA1C and CACNB2 make up approximately 11%–12% [328]. Loss of function mutations of SCN1B are responsible for BrS5. In health, this gene encodes for the β1-subunit of the inward Na^+ channel, allowing an increased Na^+ current during phase 0 of the cardiac action potential [290, 329]. Therefore, mutations of this gene lead to a volatile ion channel and hence a predisposition to arrhythmias. Mutations in the KCNE3 gene are responsible for BrS6 [330]. KCNE3 encodes for the amino acid peptide MiRP2, which is one of the homologous ancillary β-subunits of the voltage-gated K^+ channel [331]. KCNE3 mutations increase $I_{to, fast}$, inducing ST-segment elevation by aggravating transmural voltage gradients. It is hypothesized that this mutation increases the I_{to} current but also accelerates the inactivation of the current prematurely [330]. Further research does need to be conducted to establish more fully the role of KCNE3 in BrS. The SCN3B gene encodes for the β3-subunit the cardiac Na^+ channel, modifying the function of $Na_v1.5$ (Figure 44) and thereby increasing the inward Na^+ current. Brugada syndrome 7 is caused by a mutation in SCN3B, which ultimately leads to a reduction in the inward Na current [332].

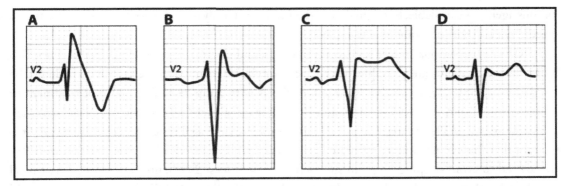

FIGURE 45: Patterns of Brugada ECG. A: The Type 1 Brugada pattern is characterized by "coved" ST-segment elevation of ≥2 mm (0.2 mV) followed by a negative T-wave. B and C: The Type 2 Brugada pattern shows a "saddleback" morphology with high takeoff (≥2 mm) ST-segment elevation, remaining ≥1 mm above the baseline, followed by a biphasic (B) or positive (C) T-wave. D: The Type 3 pattern has either a coved or saddleback morphology, with J point elevation of ≥2 mm but the terminal portion of ST-segment at <1 mm. Taken from Corrado et al. [82] with permission.

Many patients with BrS are asymptomatic until they present with cardiac arrest, although some patients may have warning symptoms such as palpitation or syncope. Some individuals may be identified through screening, due to a family history of the condition or a first-degree relative who has died from SADS. In patients with BrS, clinical examination is usually normal but still important to exclude other cardiac conditions. Meticulous ECG evaluation is vital as some of findings seen in BrS are subtle. There are 3 main ECG patterns that may be observed in BrS (Figure 45), but Type 1 pattern is seen as the "classical type" (Figure 43; Figure 45A), and it is associated with the highest risk of sudden death. In the Type 1 ECG pattern, there is coved ST-segment elevation (≥2 mm) followed by an inverted T-wave seen in the right precordial leads (Figure 43; Figure 45A). In a Type 2 ECG pattern, there is a saddleback appearance with high takeoff ST-segment elevation of ≥2 mm, a trough displaying ≥1 mm ST-segment elevation and then a positive or biphasic T-wave (Figure 45B and C). In a Type 3 pattern, there is either a saddleback or coved appearance with ST-segment elevation of <1 mm (Figure 45D). A Type 1 pattern is diagnostic of BrS; however, Type 2 and 3 patterns are seen in other conditions such as acute anterior myocardial infarction, pericarditis, and athlete's heart and may be secondary to drug therapy, so are therefore less diagnostic [317]. As previously mentioned, these ECG morphologies can often be hidden and may only be unmasked by provocation or other stimuli such as stress states, certain drugs (particularly Na$^+$ channel blocking agent), or electrolyte disturbance [317]. When the diagnosis is suspected in a patient without a Type 1 ECG pattern, pharmacological provocation testing with an Na$^+$ channel blocker is used.

Various agents are available, but Ajmaline is the most commonly used due to its high sensitivity and specificity and short duration of action [333].

Given that various factors such as fever, drugs, and electrolyte disturbance can unmask BrS ECG changes, patients are given lifestyle advice to minimize their risk of SCD. Advice includes the avoidance of hot baths, aggressive treatment of fevers with antipyrexics, and prompt treatment of conditions predisposing to dehydration and electrolyte imbalance (particularly hypokalaemia) such as gastroenteritis [333]. As ingestion of a large meal increases vagal tone and slows heart rate, it is thought that this can precipitate arrhythmias in those with BrS [334]. Therefore, patients with BrS are advised to avoid sleeping on a full stomach, leaving at least 2–3 hours between their last meal and the time they go to sleep. Several drugs can induce the Brugada ECG phenotype and should be avoided in patients with BrS. These include class 1 antiarrhythmics, alpha and beta adrenergic blockers, calcium channel blockers, and tricyclic and selective serotonin reuptake inhibiting antidepressants. Cocaine and alcohol intoxication can also precipitate arrhythmias in these patients and should be avoided [333]. Given the physiological changes exercise can have on the body such as raised core temperature and resting bradycardia, all athletes are advised to avoid moderate to high intensity sports. Both the European and American consensus is that athletes can continue to play low-static, low-dynamic sports at recreational level; this applies to those with an ICD in situ but provided that there is not a risk of bodily collision [188, 279].

Currently, only ICD insertion has been proven to reduce mortality from the condition and it is indicated for patients with high-risk features [317, 335–337]. Although inducible VT at programmed ventricular stimulation is regarded by some as a marker for poor prognosis [321], this is a subject of an ongoing controversy given that other groups have not reproduced these findings [338, 339]. In addition, false positive and nonspecific responses, particularly when aggressive stimulation protocols are used, may be observed in apparently healthy individuals [340]. Therefore, electrophysiological studies are only recommended if there are associated supraventricular arrhythmias [317, 333].

4.4.4 Wolff-Parkinson-White Syndrome

Wolff–Parkinson–White (WPW) syndrome is a cardiac condition characterized by paroxysmal arrhythmias secondary to ventricular pre-excitation due to the presence of an accessory pathway. It has an estimated prevalence in the general population of 0.13% [41]. The arrhythmias caused by this condition are usually supraventricular in nature (either orthodromic or antidromic atrioventricular re-entry tachycardia or AF), but these can occasionally degenerate into life-threatening ventricular arrhythmias including ventricular fibrillation. The most common tachyarrhythmia seen in WPW syndrome is orthodromic tachycardia. In this form, the action potential is conducted from the AV node antegradely, and back up through the accessory pathway retrogradely. Around one

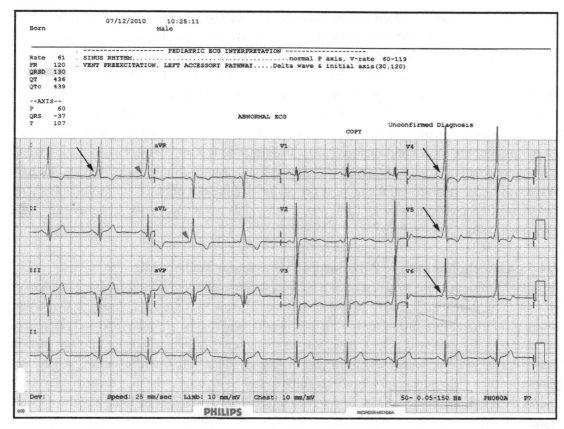

```
Born                07/12/2010    10:25:11
                                  Male


        ------------------- PEDIATRIC ECG INTERPRETATION -------------------
Rate   61   . SINUS RHYTHM.................................normal P axis, V-rate  60-119
PR    120   . VENT PREEXCITATION, LEFT ACCESSORY PATHWAY.....Delta wave & initial axis(30,120)
QRSD  130
QT    436
QTc   439

--AXIS--
P      60
QRS   -37                          ABNORMAL ECG
T     107

                                                        COPY

                                             Unconfirmed Diagnosis
```

FIGURE 46: ECG from a patient with WPW syndrome. Note the delta wave (arrows) and short PR interval in some leads (red arrowheads).

third of patients with WPW syndrome develop AF. This can be particularly dangerous if the accessory pathway conducts antegradely with a short refractory period, as the ventricles can then conduct at the same rate as the atria, which may precipitate ventricular fibrillation and SCD [188].

Patients may complain of intermittent palpitations, but occasionally, they are asymptomatic. If the diagnosis is suspected, they should undergo a full history, examination, and ECG. The characteristic ECG abnormality is a short PR interval with the presence of a delta wave indicative of pre-excitation (Figure 46). Investigation of individuals with WPW syndrome should include echocardiography as there is a rare but observed relationship between ventricular pre-excitation, ventricular hypertrophy, and other cardiac abnormalities such as Ebstein's anomaly. This association appears to have a genetic basis as Gollob et al. [341], found a PRKAG2 mutation in 2 families with ventricular pre-excitation and ventricular hypertrophy. In addition, as mentioned in section 4.1.1, certain phenocopies of HCM (such PRKAG2- and LAMP2-related cardiomyopathy) characteristically display ventricular pre-excitation on the resting ECG.

All athletes wishing to continue participating at the highest levels of physical activity should undergo risk stratification through electrophysiological studies. Indication for subsequent catheter ablation is the presence of a "high-risk" accessory pathway capable of conducting at particularly fast heart rates (i.e., with a refractory period of ≤250 msec). Athletes are able to return to their sport 3 months after successful ablation [279]. The situation is more complex when the individual is asymptomatic but is found to have a pre-excited ECG. Although they technically do not have WPW syndrome, they are still felt to be at a small but definite risk of SCD and they should therefore strongly be advised to have an EPS, with catheter ablation if a high-risk accessory pathway is discovered [188].

4.4.5   Catecholaminergic Polymorphic Ventricular Tachycardia

In 1995, Leenhardt et al. [342] described the distinct clinical entity of catecholaminergic polymorphic ventricular tachycardia (CPVT) in their series of 21 patients. Catecholaminergic polymorphic ventricular tachycardia is one of the most lethal inherited ion channel disorders and is characterized by adrenergically mediated polymorphic VT. In the absence of anti-adrenergic therapy, 30% of those affected will experience SCD by the age of 40 years [343, 344]. Catecholaminergic polymorphic ventricular tachycardia typically affects children and adolescents, who may experience syncope and/or SCA usually during exercise or catecholamine stress. There appears to be a positive family history of SCD or syncope in childhood or adolescence in around one third of patients [342].

The genetic basis for CPVT was first described by Swan et al. [345], who looked at the condition in 2 Finnish families. This group found the disorder to have an autosomal dominant inheritance. Furthermore, they linked the causative mutation to 1q42–q43 [345]. A few years later, 2 separate groups independently discovered the disease to be associated with mutations of the cardiac Ryanodine Receptor Gene (RyR2) [346, 347]. The RyR2 gene encodes for the Ca^{2+} release channel in the sarcoplasmic reticulum; in the normal heart, this Ca^{2+} release couples cross-bridge cycling [348]. It is estimated that approximately half of patients with CPVT have a RyR2 mutation [299]. In mutant RyR2 channels, there is leakage of Ca^{2+} from the sarcoplasmic reticulum into the cytoplasm [349]. This Ca^{2+} is pumped out of the cell in exchange for 3 Na^+ ions by the Na^+/Ca^{2+} pump, thereby creating an inward Na^+ current. This inward Na^+ current can lead to delayed after-depolarisations (DADs), hence predisposing to ventricular tachyatthythmias [349].

In 2002, Lahat et al. [350] investigated SCD in 7 nuclear families of a Bedouin Tribe in Israel. They discovered a new autosomal recessive variant of CPVT, which is linked to chromosome 1p13–21, caused by a mutation in the Calsequestrin (CASQ2) gene. More recently, CPVT was reported in patients who have heterozygous CASQ2 mutations [351]. Calsequestrin is a Ca^{2+}

binding protein, which acts a large reservoir of Ca^{2+} in the sarcoplasmic reticulum [352]. So far, 67 mutations of RyR2 and 7 mutations of calsequestrin have been discovered [299].

Individuals affected with CPVT have normal resting ECGs and structurally normal hearts. Although many patients are asymptomatic until they present with SCD, some may experience palpitations due to supraventricular tachycardia or AF. In an individual in whom the CPVT is suspected, the diagnosis is made on symptoms with the detection of stress-induced arrhythmias on exercise stress test or 24-hour Holter monitoring. During exercise stress testing, the individual may initially exhibit a supraventricular arrhythmia, but this may transform over the course of the investigation to VT [291]. The VT seen may be polymorphic, but bidirectional VT (where there is beat to beat 180° alternating QRS axis) is almost pathognomic of the condition, although rarely seen (Figure 47) [353].

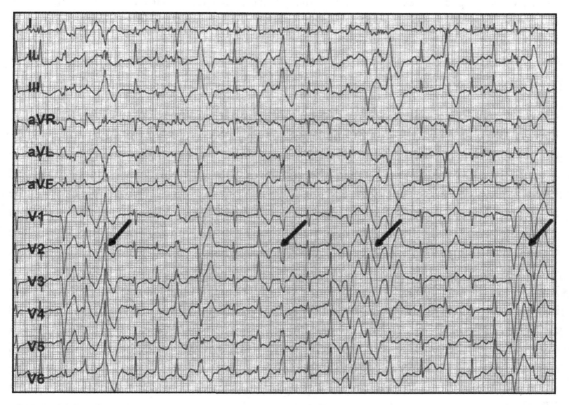

FIGURE 47: ECG from an exercise tolerance test in a patient with catecholaminergic polymorphic ventricular tachycardia. Arrowed are multiple ventricular ectopics with bidirectional couplets. Note the alternating QRS axis with 180° rotation on a beat-to-beat basis. Taken from Bastiaenen and Behr [327] with permission.

Genetic testing should be performed in individuals in whom the diagnosis is suspected. These include those who present with exercise-induced syncope or SCA and have a normal QTc. It should also be considered in those who have a genetically confirmed relative with the condition. Given the fact that arrhythmias are precipitated by adrenergic stimulation, beta-blockers are first-line therapy in symptomatic patients and also asymptomatic gene carriers [299]. The potential therapeutic use of flecainide has shown some promise in animal studies. It is hypothesized that the sodium channel blocking action of flecainide may exert an effect on the RyR2 receptor, inhibit DADs [354]. It is felt that these DADs are responsible for precipitating polymorphic and bidirectional VT. Failing these therapies, ICD insertion is the only option for those with high-risk features such as unheralded syncope and sustained VT despite beta-blockers [299].

Current guidelines for participation in sport state that symptomatic patients (who generally have a poor prognosis unless treated with an ICDrillator) should be restricted from competitive sports, with the possible exception of minimal contact, class IA activities. Patients with CPVT should also be restricted from competitive swimming. Asymptomatic patients detected as part of familial screening with documented exercise- or isoproterenol-induced VT should refrain from all competitive sports except possibly class IA activities. A less restrictive approach may be possible for the genotype-positive/phenotype-negative (i.e., asymptomatic, no inducible VT) athlete [279].

4.4.6   Early Repolarization Syndrome and Idiopathic Ventricular Fibrillation

Early repolarization (ER) is an ECG finding commonly observed in young, athletic individuals. Conventionally, it has been regarded as a benign normal variant, especially when expressed in the anterolateral ECG leads. Wasserburger, who initially described the phenomenon in 1961, defined ER as an elevated ST-segment at the end of the QRS (J point), with a downward concavity of the ST-segment and symmetrical T-waves (especially in the lateral leads). More recently, there has been growing interest in ER following the observation that it may not, in fact, be as benign as was once thought. For example, Haïssaguerre et al. [355] noted an increased prevalence of ER in patients with a history of VF [355]. This group re-defined the criteria for ER, suggesting that it is present if there is J point elevation of at least 1 mm (0.1 mV), either as QRS slurring or notching, in the inferior or lateral leads; this may or may not be in the presence of ST-segment elevation (Figures 48 and 49). When the ECGs of 206 patients who were successfully resuscitated from cardiac arrest due to idiopathic VF were studied, the presence of ER was significantly higher in the SCD group compared to controls (31% vs. 5%, P<0.001) [356].

Tikkanen et al. [357] performed a large, population-based observational study, finding that ER specifically in the inferior leads is associated with an increased risk of death from cardiac causes, a finding that is at odds with the fact that ER is commonly found in young competitive athletes (see below). The same group went on to explore this concept in further detail. They subdivided ER ECG changes into horizontal/descending ST-segments and rapidly up sloping ST-segments, the

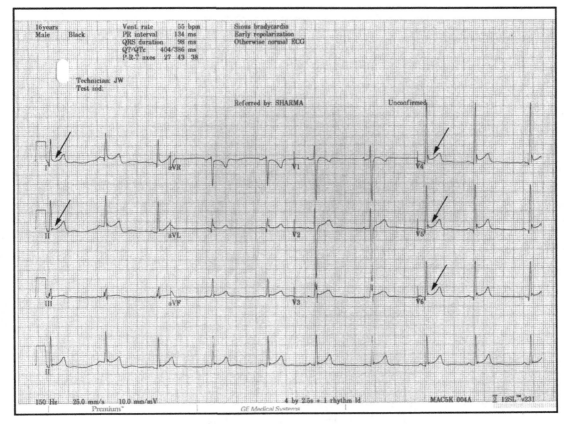

FIGURE 48: An ECG demonstrating changes of notched early repolarization (arrows).

latter pattern often being observed in young athletes and akin to Wasserburger's original definition. They found that the former was associated with a malignant outcome with SCD, whereas the latter was not. Moreover, they demonstrated that benign ER is associated with a significantly shorter QTc, whereas the malignant variant is associated with a prolongation of the QRS duration [359]. Subsequently, it has been postulated that benign ER is due to earlier repolarization, where in fact the malignant type may reflect abnormal depolarization, possibly in the presence of subtle structural heart disease [360].

Early repolarization associated with SCD, referred to "ER syndrome," has features which overlap with BrS. For this reason, the phrase 'J-wave syndromes' has been used as a blanket term for both, reflecting the fact that these entities may represent 2 spectrums of the same condition [361]. Both conditions are commoner in men, have a peak incidence of SCD in the fourth decade of life, and have a similar response to isoproterenol and quinidine [317, 356, 362]. Loss of function Ca^{2+} channel mutations are seen in both [363]; however, there does appear to be distinct differences in

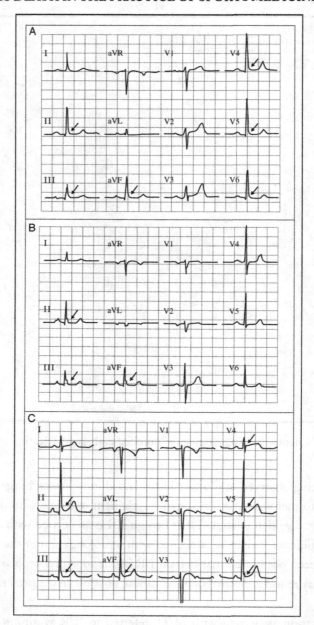

FIGURE 49: Different types of early repolarization. A: Slurred early repolarization without ST-segment elevation in the inferolateral leads. B: Notched early repolarization without ST-segment elevation in the inferior leads. C: Notched early repolarization with ST-segment elevation in the inferolateral leads; this type is consistent with Wasserberger's early description. Taken from Bastiaenen and Behr [358], with permission.

the 2 conditions. First, both differ with the effect of temperature and its impact on precipitating SCD. Sudden cardiac death in BrS is associated with pyrexia, whereas SCD in ER syndrome is precipitated by hypothermia [364]. In addition, sodium channel blockers unmask BrS ECG changes but appear to have little effect in ER syndrome [317, 356].

Since the association of ER with VF and SCD, several attempts have been made to define its prevalence in healthy cohorts and the general population. Noseworthy et al. [365] sought to determine the clinical correlates and heritability of the ER pattern in participants of the Framingham Heart Study (n=3995) and the Health 2000 Survey (n=5489), finding ER present in 6.1% of the former and 3.3% of the latter populations. The presence of ER was independently associated with male sex, younger age, lower systolic blood pressure, higher Sokolow–Lyon index, and lower Cornell voltage criteria for LVH. Looking at the Framingham population, there was evidence for a heritable basis, with siblings of individuals with ER having an ER prevalence of 11.6%.

Noseworthy et al. [366] also looked at the prevalence of ER in a young cohort of athletes (n=879, mean age 18.4 ± 0.8 years), establishing a strong association between exercise and ER. Early repolarization was associated with male gender, black ethnicity, slower heart rates, and amount of exercise performed. During a follow-up period of 21 ± 13 months, there were no cases of SCD, unexplained syncope, or hospitalisation with a cardiovascular diagnosis. Approximately 25% of the cohort demonstrated ER in the inferior or lateral leads, in accordance with previous studies [367], although the inferior subtype was comparatively uncommon. Interestingly, there was no association between the presence of ER and structural changes on echocardiography such as increased chamber size or wall thickness, implying that ER develops alongside athletic training but independently from structural remodelling and does not indicate an underlying structural abnormality.

4.4.7 Progressive Cardiac Conduction Defect

Progressive cardiac conduction defect (PCCD), also known as Lenègre [368] or Lev disease [369, 370], is a rare condition characterized by progressive alteration of the cardiac conduction tissue leading to complete atrioventricular block, syncope, and sudden cardiac death. Ventricular arrhythmias have also been reported [371]. It is considered an exaggerated ageing process, resulting in age-related sclerosis affecting the cardiac conduction tissue alone (His–Purkinje system) [372]. As of yet, the only gene associated with familial forms of the condition and causing an isolated cardiac conduction defect is SCN5A, which encodes for the α-subunit of voltage-gated cardiac Na^+ channels [372]. At present, 11 SCN5A mutations have been reported, the majority of which are associated with an autosomal dominant inheritance pattern and result in a loss of function [371]. There is a fascinating overlap between PCCD and BrS; some families are described in which the same G1408R SCN5A mutation leads to either BrS or PCCD, depending on the familial branches [373]. In addition, it has been demonstrated that BrS patients with SCN5A mutations have impaired cardiac conduction compared to patients with BrS not associated with SCN5A [377].

TABLE 13: Recommendations for bradycardia and AV block. Modified from Pelliccia et al. [188] with permission.

LESION	CRITERIA FOR ELIGIBILITY	RECOMMENDATIONS	FOLLOW-UP
Marked sinus bradycardia (<40 b.p.m.) and/or sinus pauses ≥3 s with symptoms	a) If symptoms[a] are present b) After >3 months from resolution of symptoms[a], off therapy	a) Temporary interruption of sport b) All sports	Yearly
a) AV block first and second degree, type 1 b) AV block second degree, type 2 or advanced	a) If no symptoms[a], no cardiac disease, with resolution during exercise b) In the absence of symptoms, cardiac disease, ventricular arrhythmias during exercise, and if resting heart rate is >40 b.p.m.	a) All sports b) Low-moderate dynamic, low-moderate static sports (I A,B + II A, B)	Yearly

For athletes with structural heart disease, see the recommendations of the disease.
ECG, 12-lead electrocardiogram; Echo, echocardiography; ET, exercise testing; 24 h Holter, 24 h Holter monitoring; EP, electrophysiologic; Sport types, see Table 1.
[a]Symptoms include pre-syncope, lightheadedness, exertional fatigue

Treatment involves insertion of a permanent pacemaker. The rules for athletic participation mirror those for other causes of complete AV block, as set out in Table 13 [188].

4.5   ACQUIRED CAUSES

The most common acquired causes of SCD in young athletes are commotio cordis and myocarditis.

4.5.1   Commotio Cordis

Commotio cordis is SCD from VF as a result of a blunt trauma to the chest wall. Death is usually instantaneous and survival only around 15%, only being possible if there is prompt and immediate defibrillation. The precise incidence is unknown, but commotio cordis appears to be higher in children and adolescents due to the highly compliant and usually thin chest wall in these individuals. Commotio cordis is associated mostly with sports, which utilize projectile objects, or contact sports in which forceful impacts to the chest wall can occur, such as baseball, hockey, martial arts, and ice hockey. Animal studies suggest that the trauma needs to occur at a vulnerable period of the cardiac cycle, corresponding to a narrow window of 10–20 msec on the upstroke of the T-wave, just before its peak, in order to result in sudden death [375].

Several primary prevention measures have been suggested in an attempt to reduce deaths from commotio cordis, the simplest of which is public education and awareness regarding the importance of avoiding precordial blows, even those which appear modest, for example, through improved coaching techniques [376, 377]. Other measures include improved design of sports equipment, for example, safety baseballs, and the use of chest protectors, both of which have been found to be effective at reducing deaths [378–380], although this is an area of active research given that fatalities have still occurred with both [376]. Secondary preventative measures are focused on the implementation and use of AEDs in public arenas and sports facilities, which have substantial life-saving potential and have been proven effective in terminating life-threatening ventricular tachyarrhythmias and restoring sinus rhythm [381, 382]. Automated external defibrillators are discussed in greater detail in section 7.

4.5.2   Myocarditis

Myocarditis, which is usually due to a viral illness, accounts for 7% of SCD in athletes. Inflammation and focal necrosis result in an unstable substrate predisposing to ventricular arrhythmias. Patients usually complain of coryzal symptoms, but they may be asymptomatic until they present with sudden death; this is particularly true in the athlete [41]. When identified, affected individuals should abstain from high-intensity exercise for 6 months [188, 383].

CHAPTER 5

Evaluation and Management of an Athlete

Evaluation of an athlete may be triggered through a cardiovascular screening program, implemented to exclude potentially lethal congenital or inherited cardiac disorders prior to entry into competition. However, individuals may also present to a general physician through symptoms or a positive family history of a hereditary cardiac disorder in a first-degree relative. In either case, evaluation of an athlete follows the same basic principles of history, examination, and investigations. Given the unique nature of these individuals, in many cases, referral to a specialist center with expertise in the conditions responsible for SCD death and experience in differentiating physiological adaptation from cardiac pathology is appropriate.

5.1   HISTORY

Although the majority of athletes who harbor sinister cardiac disorders are asymptomatic, the importance of a detailed and focused history cannot be underestimated. Certain "red flag" symptoms can be associated with particular conditions (Tables 14 and 15) and are therefore important to inquire about. These include exertional chest pain, syncope (with emphasis on the trigger and situation at the time of the event), palpitation, and breathlessness disproportionate to the amount of exercise being performed.

Given the genetic nature of most conditions implicated in SCD in young athletes, a detailed family history is crucial. This includes not only determining whether specific conditions are present in other family members but also inquiring about sinister symptoms and cases of premature SCD. For example, features indicative of a ventricular arrhythmia may include epilepsy, syncope, and unexplained drowning or road traffic accidents. Where a family member has been affected with premature SCD, the post-mortem report is invaluable in differentiating between congenital (e.g., anomalous coronary arteries) and hereditary (e.g., HCM) disorders, thus aiding further assessment of surviving relatives.

TABLE 14: "Red-flag" features in an athlete which should prompt referral for specialist investigation.

"RED-FLAG" FEATURES IN THE HISTORY REQUIRING SPECIALIST REFERRAL
• Exertional chest pain or discomfort
• Breathlessness disproportionate to the amount of exercise being perfumed, unexplained breathlessness or fatigue related to exercise
• Palpitation
• Syncope, or near-syncope, particular if occurring during exercise
• History of elevated blood pressure
• Prior recognition of a heart murmur
• Personal or family history of unexplained drowning or road traffic accidents
• Family history of SCD

TABLE 15: Triggers causing syncope and sudden death in the ion channel disordered LQTS and CPVT, highlighting the importance of a detailed history.

DISORDER	TRIGGER	CLINICAL EVENT
LQTS-1	Emotional stress, physical exercise, swimming, and diving into water	Syncope, sudden death, seizure, drowning or near drowning, motor vehicle accident
LQTS-2	Emotional stress, physical exercise, loud noises	Syncope, sudden death, seizure, motor vehicle accident
LQTS-3	Rest or sleep	Sudden death or sudden infant death
CPVT	Emotional stress, physical exercise	Syncope, sudden death, seizure, drowning or near drowning
LQTS = Long-QT syndrome; CPVT = catecholaminergic polymorphic ventricular tachycardia		

5.2   PHYSICAL EXAMINATION

As with symptoms, examination may be normal in an athlete. Nevertheless, several important conditions associated with SCD can produce physical signs, including HCM, mitral valve prolapse, aortic stenosis, Marfan syndrome (Figure 33; Table 16), and familial hypercholesterolemia (Table 16).

5.3   INVESTIGATIONS

Two basic investigations form the cornerstone of athletic clinical evaluation: the 12-lead electrocardiogram and transthoracic echocardiography.

5.3.1   12-Lead Electrocardiogram

Although much debate surrounds the routine use of 12-lead electrocardiography in pre-participation screening (see section 6), it is essential in evaluating an athlete with sinister symptoms or a positive family history. Several inherited and congenital conditions may be diagnosed on the ECG alone,

TABLE 16: Conditions associated with sudden cardiac death in an athlete and the physical signs they produce.

CONDITION	CLINICAL FEATURES
Hypertrophic Cardiomyopathy	Ejection systolic murmur, accentuated with standing and Valsalva maneuver, and diminished with squatting. If dynamic outflow obstruction, pulsus bisferiens and double apical impulse may be detected
Mitral Valve Prolapse	Mid-systolic click with late systolic murmur, accentuated with standing, Valsalva maneuver and hand grip, and diminished with squatting
Aortic Stenosis	Ejection systolic murmur; in severe cases, narrow pulse pressure, slow rising pulse and diminished or absent 2nd heart sound
Marfan Syndrome	Tall height, arachnodactyly, arm span > height, pectus excavatum or carinatum, high arched palate, joint hypermobility joint, lens subluxation, mitral or aortic regurgitation
Familial hypercholesterolemia	Eruptive xanthoma, xanthelasma, premature corneal arcus

including LQTS, WPW syndrome, and BrS. In other disorders (such as the primary cardiomyopathies), the first phenotypic manifestation may be on the resting ECG, prompting further evaluation. A discussion of the ECG changes that athletes may develop in response to cardiac adaptation to exercise, as well as the overlap between physiological changes and pathology in different cohorts, has already been given in section 3.3.1.

5.3.2   Echocardiography

Echocardiography, although adding to the cost of screening by up to 4.5-fold [384], is the gold standard investigation for valvular heart disease and HCM. Echocardiography may also be useful in the diagnosis and assessment of DCM and can detect CCAA in young, thin individuals with good short-axis views, thus prompting further evaluation with cardiac CMRI or CT. However, assessment of the right ventricle can prove challenging, making echocardiography of relatively less value in conditions such as the concealed phase of ARVC. In addition, the LV apex may be poorly visualized in some individuals, with the administration of contrast agents helping in this respect.

Athletic training may produce phenotypic features on echocardiography that overlap with changes observed in the primary cardiomyopathies, and a discussion of this aspect has been provided in section 3.3.2.

5.3.3   Further Investigations

Some athletes will invariably require further investigations to look for the broader phenotype of a cardiomyopathy or ion channel disorder, to image specific regions of the heart in greater detail, or for risk stratification purposes. These include signal-averaged ECG, exercise ECG, 24-hour Holter monitoring, CPEX, CMRI, and MRI or CT coronary angiography. Rarely, more invasive tests (such as coronary angiography and electrophysiological studies) may be warranted. The usefulness of genetic testing is limited by the marked heterogeneity and incomplete penetrance of conditions responsible for SCD, and combined with incomplete knowledge of all disease causing mutations, these factors prevent timely diagnosis. However, in some difficult or borderline cases, genetic testing may prove useful in resolving a clinical dilemma as previously discussed in the sections above.

5.4   MANAGEMENT

Management of an athlete is dependent on the specific condition they harbor, and this has already been described alongside the relevant conditions in section 4. In general, current guidelines are conservative and in most cases, abstinence from moderate to strenuous physical exertion and competitive sport is recommended [187, 188]. However with certain conditions amenable to curative procedures (such as WPW syndrome/pre-excitation and anomalous coronary arteries), competitive sport may be resumed after the procedure provided that specific criteria are met [187, 188].

In reality, the sports physician can face challenging management decisions when dealing with an athlete affected with a condition placing them at risk of SCD. It is often difficult to persuade a motivated, high-profile athlete to refrain from the competitive sport on which their future career and livelihood depends, and some accept the risk and continue. Several high-profile cases, including the "Hank Gathers affair," have highlighted these challenges [13].

·   ·   ·   ·

CHAPTER 6

Pre-Participation Screening

The sudden death of a young athlete is an emotional and tragic event, affecting not only the individual's family and friends but also attracting much media attention given the status of athletes as the epitome of health. Invariably, questions are asked as to which interventions could have prevented the death through pre-participation screening. Yet, the prevalence of conditions responsible for SCD in young athletes (Table 17) and the incidence of cardiovascular deaths themselves (Table 1) are low, raising concerns about the cost-effectiveness of pre-participation screening programs (PPSPs). Given these issues, although several sporting bodies including Fédération Internationale de Football Association (FIFA) and the International Olympic Committee (IOC) have recommended pre-participation cardiac evaluation; the majority of countries do not offer state-sponsored programs to athletes.

TABLE 17: Prevalence of cardiovascular disorders in athletes that put them at risk of sudden cardiac death.

STUDY	POPULATION	PREVALENCE
Baggish et al. [385], 2010	510 college athletes (US)	0.6%
Bessem et al. [386], 2009	428 athletes; age 12–35 (Netherlands)	0.7%
Hevia et al. [388], 2011	1220 amateur athletes (Spain)	0.16%
Wilson et al. [388], 2008	2720 athletes and children; age 10–17 years (UK)	0.3%
Corrado et al. [37], 2006	42,386 athletes; age 12–35 (Italy)	0.2%
Fuller et al. [389], 1997	5617 high school athletes (US)	0.4%
Viskin [340], 2003	Estimate in competitive athletes; age 12–35 (US)	0.3%

The aim of screening is to detect cardiovascular disorders that put an athlete at risk of sudden death during sport. Although most would agree that such aims are laudable, the precise method by which screening should occur is still hotly contested. In the United States, a long-standing PPSP has been in existence, which is limited to history taking and physical examination. However, there are considerable data that many of the disorders responsible for SCD in athletes are silent, and therefore, the majority of athletes who die are asymptomatic, with physical examination rarely being helpful [10, 31, 46]. Even with a focused history and examination, PPSPs based on this strategy in isolation detect <10% of individuals with a condition placing them at risk of SCD during sport [39, 338, 341]: no study exists that demonstrates that such a strategy is effective in preventing or detecting athletes at risk for SCD.

Although still highly controversial, data show that the addition of a routine 12-lead electrocardiogram to PPSPs can greatly improve the sensitivity for detecting underlying cardiac abnormalities, from anywhere between 50% and 70% in elite athletes [84, 339, 342]. There is also considerable evidence from Italy, where a mandatory state-sponsored PPSP has been operating for >25 years [121], that routine electrocardiography can reduce mortality in athletes by up to 90% when used alongside a medical history and physical examination [37], predominantly through the detection of cardiomyopathies. The success of the Italian program has led many physicians and sporting bodies to recommend routine electrocardiography in PPSPs [343–345].

The increase in sensitivity afforded by electrocardiography can, however, compromise specificity given that considerable overlap exists between physiological training-related ECG changes and those found in pathological conditions predisposing athletes to SCD. False positive rates of up to 40% have been recorded in the past [84, 335, 342], although it is becoming clear that this was likely due to the inclusion of common training-related changes (in particular voltage criteria for LVH) as abnormalities [346]. With refinement of the ECG criteria used to interpret an athlete's ECG [82], false positive rates can be lowered to a much more acceptable 10%, at least in predominantly adult Caucasian cohorts [346]. Future work aims to improve this even further as our knowledge of the electrical manifestation of the athlete's heart increases.

Antagonists of the 12-lead ECG argue that not only does it add considerably to the cost of a screening program, it cannot also detect some important causes of SCD such as aortic dilatation secondary to collagen disorders, premature coronary artery disease, and anomalous coronary arteries [347]. In addition, other studies utilizing the 12-lead ECG in PPSPs outside of Italy have not shown a reduction in SCD in athletes [348, 349]. Given these conflicting data and unresolved issues, in reality, a pragmatic approach is to raise awareness among athletes and sporting bodies about the presenting symptoms of disorders causing SCD and their familial inheritance, alongside PPSPs and strategies in effectively managing sudden cardiac arrest.

• • • • •

CHAPTER 7

Prevention and the Role of Automated External Defibrillators

As many of the conditions that cause SCD in young athletes may not be detected before the onset of cardiac arrest, many sports venues have automated external defibrillators (AEDs) on site. AEDs are computerized devices, which are able to deliver a shock to victims of cardiac arrest who have a shockable rhythm, that is, ventricular fibrillation or pulseless ventricular tachycardia. AEDs have been designed to be suitable for use by both lay people and healthcare professionals. When being operated by a lay person, careful visual and vocal prompts are given to guide rescuers. AEDs are sophisticated as they recognize the ECG rhythm and advise the user to deliver a shock if necessary. Automatic AEDs interpret the rhythm and automatically shock if there is a shockable rhythm [350].

In terms of outcome, the most important factor in survival of cardiac arrest is the length of time from cardiac arrest to defibrillation [351]. Therefore, it is crucial that the cardiac arrest is recognized as soon as possible to minimize the time to defibrillation. This can be achieved through basic life support training in the general public. Basic life support training is essential for those who work closely with athletes. As many of the cardiac conditions that cause SCA in the athlete precipitate sudden ventricular fibrillation, rapid defibrillation is the only way to abort SCD. Therefore, appreciation that features such as seizure-like activity may be the first sign of imminent cardiac arrest is vital. It is estimated that 50% of SCD in young athletes is heralded by myoclonic jerks or seizure-like activity [21]. These features can be misinterpreted to be much more benign than they are, meaning there may be a delay in adequate resuscitation attempts. Providing defibrillation occurs within the first minute, survival from ventricular fibrillation can be as high as 90% [352]. Survival rates diminish by 7%–10% for every minute delay, with an estimated survival of 2%–5% beyond 12 minutes [353].

Despite the increase in AEDs in sports grounds, the outcome of cardiac arrests in athletes is still poor. Drezner and Rogers, reported a survival rate of only 11% in their review of cardiac arrest in 9 intercollegiate athletes, despite the cardiac arrest being witnessed and CPR being started promptly. Mean time to defibrillation was 3.1 minutes [21]. A more recent study by the same group however suggests a more promising outcome. In this cohort of 1710 US high schools, which had onsite AEDs, there were 36 cases of sudden cardiac arrest. Fourteen of these victims were high school athletes with a mean age of 16 years and 22 were older non-students with a mean age of 57 years. Of these, 35 were witnessed, 34 received bystander CPR, and 30 received an AED shock. Mean time from cardiac arrest to AED shock was 3.6 minutes. There was an overall survival to hospital discharge of 64% [28].

The use of emergency response planning has been advocated by bodies such as the European Society of Cardiology. As the first responder during SCA may vary depending on who is present at the time of the event, it is advised that school and sporting clubs have an ERP. As the coach and the team physiotherapist are usually the first on the scene, it is strongly recommended that they are aware of the ERP and trained in basic life support. In large facilities, like sporting arenas or airports, multiple AEDs should be available, aiming for a collapse to defibrillation target of 3–5 minutes [351]. In sporting events such as marathons or triathlons, mobile rescue teams need to be available, which can access the victim by bicycle or car.

· · · ·

CHAPTER 8

Summary

Sudden cardiac death is the commonest cause of non-traumatic mortality in young athletes, with new studies suggesting the incidence to be higher than previously estimated. Despite the health benefits of exercise being well established, physical activity is usually the trigger for these tragic events in a small minority of athletes harboring one of several inherited or congenital cardiac conditions. Etiology appears to vary by region, with cardiomyopathies (in particular HCM and ARVC) and congenital anatomic abnormalities being among the commonest causes. Recent advances in our understanding of both the genetic and clinical mechanisms underlying these disorders have led to the development of guidelines, therapeutic interventions, and lifestyle modifications, all of which aim to aid detection, minimize risk of SCD, and direct management. Given that athletic training results in cardiac adaptive responses, which on routine investigations occasionally overlap with those observed in pathological conditions, athletes should be evaluated by physicians with experience and expertise in dealing with this unique group of individuals. Management of an athlete depends upon the specific condition they harbor, but in general, guidelines are conservative and limit competitive sport at the highest level unless a curative procedure can be undertaken. Identification of athletes at risk of SCD is challenging given that the majority of individuals with underlying disorders are asymptomatic. Although PPSPs have been advocated in an attempt to identify these individuals, debate remains as to the most cost-effective method. Given that no screening method will offer absolute protection from SCD, PPSPs must, however, be implemented alongside other strategies including education and training of the general public and athletic communities and effective emergency response plans. Automated external defibrillators are emerging as an important aspect of secondary prevention given that prompt defibrillation is the only effective treatment for SCA.

· · · ·

References

[1]   Fletcher G, Blair S, Blumenthal J, et al. Statement on exercise. Benefits and recommendations for physical activity programs for all Americans. A statement for health professionals by the Committee on Exercise and Cardiac Rehabilitation of the Council on Clinical Cardiology, American Heart Association. *Circulation.* 1992;86(1): pp. 340–4.

[2]   Morris C, Froelicher V. Cardiovascular benefits of physical activity. *Herz.* 1991;16(4): pp. 222–36.

[3]   Paffenbarger RS, Hyde RT, Wing AL, Hsieh CC. Physical activity, all-cause mortality, and longevity of college alumni. *The New England Journal of Medicine.* 1986;314(10): pp. 605–13.

[4]   Link MS, Mark Estes NA. Sudden cardiac death in athletes. *Progress in Cardiovascular Diseases.* 2008;51(1): pp. 44–57.

[5]   Bille K, Figueiras D, Schamasch P, et al. Sudden cardiac death in athletes: the Lausanne Recommendations. *European Journal of Cardiovascular Prevention and Rehabilitation.* 2006; 13(6): pp. 859–75.

[6]   Cross BJ, Estes NAM, Link MS. Sudden cardiac death in young athletes and nonathletes. *Current Opinion in Critical Care.* 2011;17(4): pp. 328–34.

[7]   Basavarajaiah S, Shah A, Sharma S. Sudden cardiac death in young athletes. *Heart.* 2007;93(3): pp. 287–9.

[8]   Lopshire JC, Zipes DP. Sudden cardiac death: better understanding of risks, mechanisms, and treatment. *Circulation.* 2006;114(11): pp. 1134–6.

[9]   Solberg EE, Gjertsen F, Haugstad E, Kolsrud L. Sudden death in sports among young adults in Norway. *European Journal of Cardiovascular Prevention and Rehabilitation Official Journal of the European Society of Cardiology Working Groups on Epidemiology Prevention and Cardiac Rehabilitation and Exercise Physiology.* 2010;17(3): pp. 337–41.

[10]   Maron BJ, Shirani J, Poliac LC, et al. Sudden death in young competitive athletes. Clinical, demographic, and pathological profiles. *Journal of the American Medical Association.* 1996;276(3): pp. 199–204.

[11] Corrado D, Thiene G, Nava A, Pennelli N, Rossi L. Sudden death in young competitive athletes: clinicopathologic correlations in 22 cases. *American Journal of Medicine*. 1990;89(5): pp. 588–96.

[12] Maron BJ, Doerer JJ, Haas TS, Tierney DM, Mueller FO. Sudden deaths in young competitive athletes: analysis of 1866 deaths in the United States, 1980–2006. *Circulation*. 2009;119(8): pp. 1085–92.

[13] Maron B. Sudden death in young athletes. Lessons from the Hank Gathers affair. *New England Journal of Medicine*. 1993;329(1): pp. 55–7.

[14] Mosterd A, Senden JP, Engelfriet P. Preventing sudden cardiac death in athletes: finding the needle in the haystack or closing the barn door? *European Journal of Cardiovascular Prevention and Rehabilitation Official Journal of the European Society of Cardiology Working Groups on Epidemiology Prevention and Cardiac Rehabilitation and Exercise Physiology*. 2011;18(2): pp. 194–6.

[15] Rao AL, Standaert CJ, Drezner JA, Herring SA. Expert opinion and controversies in musculoskeletal and sports medicine: preventing sudden cardiac death in young athletes. *Archives of Physical Medicine and Rehabilitation*. 2010;91(6): pp. 958–62.

[16] Corrado D, Migliore F, Bevilacqua M, Basso C, Thiene G. Sudden cardiac death in athletes: can it be prevented by screening? *Herz*. 2009;34(4): pp. 259–66.

[17] Harmon KG, Asif IM, Klossner D, Drezner JA. Incidence of sudden cardiac death in national collegiate athletic association athletes. *Circulation*. 2011;123(15): pp. 1594–600.

[18] Eckart RE, Scoville SL, Campbell CL, et al. Sudden death in young adults: a 25-year review of autopsies in military recruits. *Annals of Internal Medicine*. 2004;141(11): pp. 882–4.

[19] Van Camp S, Bloor C, Mueller F, Cantu R, Olson H. Nontraumatic sports death in high school and college athletes. *Medicine and Science in Sports and Exercise*. 1995;27(5): pp. 641–7.

[20] Montagnana M, Lippi G, Franchini M, Banfi G, Guidi GC. Sudden cardiac death in young athletes. *Internal Medicine*. 2008;47(15): pp. 1373–8.

[21] Drezner JA, Rogers KJ. Sudden cardiac arrest in intercollegiate athletes: detailed analysis and outcomes of resuscitation in nine cases. *Heart Rhythm*. 2006;3(7): pp. 755–9.

[22] Noakes TD, Opie LH, Rose AG, et al. Autopsy-proved coronary atherosclerosis in marathon runners. *The New England Journal of Medicine*. 1979;301(2): pp. 86–9.

[23] Thompson PD, Funk EJ, Carleton RA, Sturner WQ. Incidence of death during jogging in Rhode Island from 1975 through 1980. *Journal of the American Medical Association*. 1982;247(18): pp. 2535–8.

[24] Waller BF, Roberts WC. Sudden death while running in conditioned runners aged 40 years or over. *The American Journal of Cardiology*. 1980;45(6): pp. 1292–300.

[25] Corrado D, Basso C, Schiavon M, Pelliccia A, Thiene G. Pre-participation screening of young competitive athletes for prevention of sudden cardiac death. *Journal of the American College of Cardiology.* 2008;52(24): pp. 1981–9.

[26] Corrado D, Pelliccia A, Bjørnstad HH, et al. Cardiovascular pre-participation screening of young competitive athletes for prevention of sudden death: proposal for a common European protocol. *European Heart Journal.* 2005;26(5): pp. 516–24.

[27] De Ceuninck M, D'Hooghe M, D'Hooghe P. Sudden cardiac death in football. *FIFA and UEFA Sports Medical Committee.* 2005. Available at: http://eur.i1.yimg.com/eur.yimg.com/i/eu/fifa/do/commi.pdf.

[28] Drezner JA, Rao AL, Heistand J, Bloomingdale MK, Harmon KG. Effectiveness of emergency response planning for sudden cardiac arrest in United States high schools with automated external defibrillators. *Circulation.* 2009;120(6): pp. 518–25.

[29] Drezner JA, Courson RW, Roberts WO, et al. Inter-association task force recommendations on emergency preparedness and management of sudden cardiac arrest in high school and college athletic programs: a consensus statement. *Journal of Athletic Training.* 2007;42(1): pp. 143–58.

[30] Drezner JA. Preparing for sudden cardiac arrest—the essential role of automated external defibrillators in athletic medicine: a critical review. *British Journal of Sports Medicine.* 2009;43(9): pp. 702–7.

[31] Maron BJ. Sudden death in young athletes. *New England Journal of Medicine.* 2003;349: 1064–75.

[32] Corrado D, Basso C, Rizzoli G, Schiavon M, Thiene G. Does sports activity enhance the risk of sudden death in adolescents and young adults? *Journal of the American College of Cardiology.* 2003;42(11): pp. 1959–63.

[33] Maron BJ. Cardiovascular risks to young persons on the athletic field. *Annals of Internal Medicine.* 1998;129(5): pp. 379–86.

[34] Scoville SL, Gardner JW, Magill AJ, Potter RN, Kark JA. Nontraumatic deaths during U.S. Armed Forces basic training, 1977–2001. *American Journal of Preventive Medicine.* 2004;26(3): pp. 205–12.

[35] Maron BJ, Carney KP, Lever HM, et al. Relationship of race to sudden cardiac death in competitive athletes with hypertrophic cardiomyopathy. *Journal of the American College of Cardiology.* 2003;41(6): pp. 974–80.

[36] Atkins DL, Everson-Stewart S, Sears GK, et al. Epidemiology and outcomes from out-of-hospital cardiac arrest in children: the Resuscitation Outcomes Consortium Epistry-Cardiac Arrest. *Circulation.* 2009;119(11): pp. 1484–91.

[37]   Corrado D, Basso C, Pavei A, et al. Trends in sudden cardiovascular death in young competitive athletes after implementation of a preparticipation screening program. *Journal of the American Medical Association*. 2006;296(13): pp. 1593–601.

[38]   Drezner JA, Rogers KJ, Zimmer RR, Sennett BJ. Use of automated external defibrillators at NCAA Division I universities. *Medicine & Science in Sports & Exercise*. 2005;37(9): pp. 1487–92.

[39]   Maron BJ, Gohman TE, Aeppli D. Prevalence of sudden cardiac death during competitive sports activities in Minnesota high school athletes. *Journal of the American College of Cardiology*. 1998;32(7): pp. 1881–4.

[40]   Chugh SS, Reinier K, Balaji S, et al. Population-based analysis of sudden death in children: the Oregon Sudden Unexpected Death Study. *Heart Rhythm*. 2009;6(11): pp. 1618–22.

[41]   Sheikh N, Sharma S. Overview of sudden cardiac death in young athletes. *The Physician and Sports Medicine*. 2011;39(4): pp. 22–36.

[42]   Opthof T. The normal range and determinants of the intrinsic heart rate in man. *Cardiovascular Research*. 2000;45(1): pp. 492–501.

[43]   Uusitalo AL, Uusitalo AJ, Rusko HK. Exhaustive endurance training for 6–9 weeks did not induce changes in intrinsic heart rate and cardiac autonomic modulation in female athletes. 1998: pp. 532–40.

[44]   Rowell L. *Human Circulation: Regulation During Physical Stress*. New York, NY: Oxford University Press; 1986.

[45]   Pelliccia A, Maron BJ, Spataro A, Proschan MA, Spirito P. The upper limit of physiologic cardiac hypertrophy in highly trained elite athletes. *The New England Journal of Medicine*. 1991;324(5): pp. 295–301.

[46]   Pelliccia A, Culasso F, Di Paolo FM, Maron BJ. Physiologic left ventricular cavity dilatation in elite athletes. *Annals of Internal Medicine*. 1999;130(1): pp. 23–31.

[47]   Pelliccia A, Maron BJ, Culasso F, Spataro A, Caselli G. Athlete's heart in women—echocardiographic characterization of highly trained elite female athletes. *Journal of the American Medical Association*. 1996;276(3): pp. 211–5.

[48]   Makan J, Sharma S, Firoozi S, et al. Physiological upper limits of ventricular cavity size in highly trained adolescent athletes. *Heart (British Cardiac Society)*. 2005;91(4): pp. 495–9.

[49]   Sharma S, Maron BJ, Whyte G, et al. Physiologic limits of left ventricular hypertrophy in elite junior athletes: relevance to differential diagnosis of athlete's heart and hypertrophic cardiomyopathy. *Journal of the American College of Cardiology*. 2002;40(8): pp. 1431–6.

[50]   Rawlins J, Carre F, Kervio G, et al. Ethnic differences in physiological cardiac adaptation to intense physical exercise in highly trained female athletes. *Circulation*. 2010;121(9): pp. 1078–85.

[51] Magalski A, Maron BJ, Main ML, et al. Relation of race to electrocardiographic patterns in elite American football players. *Journal of the American College of Cardiology*. 2008;51(23): pp. 2250–5.

[52] Basavarajaiah S, Boraita A, Whyte G, et al. Ethnic differences in left ventricular remodeling in highly-trained athletes relevance to differentiating physiologic left ventricular hypertrophy from hypertrophic cardiomyopathy. *Journal of the American College of Cardiology*. 2008;51(23): pp. 2256–62.

[53] Papadakis M, Carre F, Kervio G, et al. The prevalence, distribution, and clinical outcomes of electrocardiographic repolarization patterns in male athletes of African/Afro-Caribbean origin. *European Heart Journal*. 2011;32(18): pp. 2304–13.

[54] Montgomery HE, Clarkson P, Dollery CM, et al. Association of angiotensin-converting enzyme gene I/D polymorphism with change in left ventricular mass in response to physical training. 1997: pp. 741–7.

[55] Diet F, Graf C, Mahnke N, et al. ACE and angiotensinogen gene genotypes and left ventricular mass in athletes. *European Journal of Clinical Investigation*. 2001;31(10): pp. 836–42.

[56] Boraita A, De La Rosa A, Heras ME, et al. Cardiovascular adaptation, functional capacity and Angiotensin-converting enzyme I/D polymorphism in elite athletes. *Revista Española de Cardiología*. 2010;63(7): pp. 810–9.

[57] Rizzo M, Gensini F, Fatini C, et al. ACE I/D polymorphism and cardiac adaptations in adolescent athletes. *Medicine & Science in Sports & Exercise*. 2003;35(12): pp. 1986–90.

[58] Nagashima J, Musha H, Takada H, et al. Influence of angiotensin-converting enzyme gene polymorphism on development of athlete's heart. *Clinical Cardiology*. 2002;23(1): pp. 621–4.

[59] Di Mauro M, Izzicupo P, Santarelli F, et al. ACE and AGTR1 polymorphisms and left ventricular hypertrophy in endurance athletes. *Medicine & Science in Sports & Exercise*. 2010;42(5): pp. 915–21.

[60] Hernández D, De La Rosa A, Barragán A, et al. The ACE/DD genotype is associated with the extent of exercise-induced left ventricular growth in endurance athletes. *Journal of the American College of Cardiology*. 2003;42(3): pp. 527–32.

[61] Montgomery H, Brull D, Humphries SE. Analysis of gene-environment interactions by "stressing-the-genotype" studies: the angiotensin converting enzyme and exercise-induced left ventricular hypertrophy as an example. *Italian Heart Journal Official Journal of the Italian Federation of Cardiology*. 2002;3(1): pp. 10–4.

[62] Karjalainen J, Kujala UM, Stolt A, et al. Angiotensinogen gene M235T polymorphism predicts left ventricular hypertrophy in endurance athletes. *Journal of the American College of Cardiology*. 1999;34(2): pp. 494–9.

[63]   Miyata S, Haneda T. Hypertrophic growth of cultured neonatal rat heart cells mediated by type 1 angiotensin II receptor. *American Journal of Physiology*. 1994;266(6 Pt 2): pp. H2443–51.

[64]   Everett AD, Tufro-McReddie A, Fisher A, Gomez RA. Angiotensin receptor regulates cardiac hypertrophy and transforming growth factor-beta 1 expression. *Hypertension*. 1994;23(5): pp. 587–92.

[65]   Dostal DE, Rothblum KN, Chernin MI, Cooper GR, Baker KM. Intracardiac detection of angiotensinogen and renin: a localized renin-angiotensin system in neonatal rat heart. *American Journal of Physiology*. 1992;263(4 Pt 1): pp. C838–50.

[66]   Dostal DE, Rothblum KN, Conrad KM, Cooper GR, Baker KM. Detection of angiotensin I and II in cultured rat cardiac myocytes and fibroblasts. *American Journal of Physiology*. 1992;263(4 Pt 1): pp. C851–63.

[67]   Rogers TB, Gaa ST, Allen IS. Identification and characterization of functional angiotensin II receptors on cultured heart myocytes. *The Journal of Pharmacology and Experimental Therapeutics*. 1986;236(2): pp. 438–44.

[68]   Brown NJ, Blais C, Gandhi SK, Adam A. ACE insertion/deletion genotype affects bradykinin metabolism. *Journal of Cardiovascular Pharmacology*. 1998;32(3): pp. 373–7.

[69]   Ito H, Hiroe M, Hirata Y, et al. Insulin-like growth factor-II induces hypertrophy with increased expression of muscle specific genes in cultured rat cardiomyocytes. *Journal of Molecular and Cellular Cardiology*. 1994;26(7): pp. 1715–21.

[70]   Donath MY, Zapf J, Eppenberger-Eberhardt M, Froesch ER, Eppenberger HM. Insulin-like growth factor I stimulates myofibril development and decreases smooth muscle alpha-actin of adult cardiomyocytes. *Proceedings of the National Academy of Sciences of the United States of America*. 1994;91(5): pp. 1686–90.

[71]   Decker RS, Cook MG, Behnke-Barclay M, Decker ML. Some growth factors stimulate cultured adult rabbit ventricular myocyte hypertrophy in the absence of mechanical loading. *Circulation Research*. 1995;77(3): pp. 544–55.

[72]   Reiss K, Cheng W, Kajstura J, et al. Fibroblast proliferation during myocardial development in rats is regulated by IGF-1 receptors. *American Journal of Physiology*. 1995;269(3 Pt 2): pp. H943–51.

[73]   Pauliks LB, Cole KE, Mergner WJ. Increased insulin-like growth factor-1 protein in human left ventricular hypertrophy. *Experimental and Molecular Pathology*. 1999;66(1): pp. 53–8.

[74]   DeBosch B, Treskov I, Lupu TS, et al. Akt1 is required for physiological cardiac growth. *Circulation*. 2006;113(17): pp. 2097–104.

[75]   Schlüter KD, Goldberg Y, Taimor G, Schäfer M, Piper HM. Role of phosphatidylinositol

3-kinase activation in the hypertrophic growth of adult ventricular cardiomyocytes. *Cardiovascular Research.* 1998;40(1): pp. 174–81.

[76] Roeske WR, Rourke RAO, Klein A, Leopold G, Karliner JS. Noninvasive evaluation of ventricular hypertrophy in professional athletes. *Circulation.* 1976;53(2): pp. 286–91.

[77] Maron BJ. Structural features of the athlete heart as defined by echocardiography. *Journal of the American College of Cardiology.* 1986;7(1): pp. 190–203.

[78] Scharf M, Brem MH, Wilhelm M, et al. Cardiac magnetic resonance assessment of left and right ventricular morphologic and functional adaptations in professional soccer players. *American Heart Journal.* 2010;159(5): pp. 911–8.

[79] Sipola P, Magga J, Husso M, et al. Cardiac MRI assessed left ventricular hypertrophy in differentiating hypertensive heart disease from hypertrophic cardiomyopathy attributable to a sarcomeric gene mutation. *European Radiology.* 2011;21(7): pp. 1383–9.

[80] Scharhag J, Schneider G, Urhausen A, et al. Athlete's heart: right and left ventricular mass and function in male endurance athletes and untrained individuals determined by magnetic resonance imaging. *Journal of the American College of Cardiology.* 2002;40(10): pp. 1856–63.

[81] Prakken NH, Velthuis BK, Teske AJ, et al. Cardiac MRI reference values for athletes and nonathletes corrected for body surface area, training hours/week and sex. *European Journal of Cardiovascular Prevention and Rehabilitation Official Journal of the European Society of Cardiology Working Groups on Epidemiology Prevention and Cardiac Rehabilitation and Exercise Physiology.* 2010;17(2): pp. 198–203.

[82] Corrado D, Pelliccia A, Heidbuchel H, et al. Recommendations for interpretation of 12-lead electrocardiogram in the athlete. *European Heart Journal.* 2010;31(2): pp. 243–59.

[83] White PD. The Pulse After a marathon race. *Journal of the American Medical Association.* 1918;71(13): pp. 1047–8.

[84] Pelliccia A, Maron BJ, Culasso F, et al. Clinical significance of abnormal electrocardiographic patterns in trained athletes. *Circulation.* 2000;102(3): pp. 278–84.

[85] Huston TP, Puffer JC, Rodney WM. The athletic heart syndrome. *The New England Journal of Medicine.* 1985;3(2): pp. 259–72.

[86] Pelliccia A, Culasso F, Di Paolo FM, et al. Prevalence of abnormal electrocardiograms in a large, unselected population undergoing pre-participation cardiovascular screening. *European Heart Journal.* 2007;28(16): pp. 2006–10.

[87] Grimm W, Hoffmann J, Menz V, et al. Intrinsic sinus and atrioventricular node electrophysiologic adaptations in endurance athletes. *Journal of the American College of Cardiology.* 2002;77(6): pp. 1033–8.

[88]   Balady GJ, Cadigan JB, Ryan TJ. Electrocardiogram of the athlete: an analysis of 289 pro-
       fessional football players. *The American Journal of Cardiology*. 1984;53(9): pp. 1339–43.

[89]   Sharma S, Whyte G, Elliott P, et al. Electrocardiographic changes in 1000 highly trained
       junior elite athletes. *British Journal of Sports Medicine*. 1999;33(5): pp. 319–24.

[90]   Venerando A, Rulli V. Frequency morphology and meaning of the electrocardiographic
       anomalies found in Olympic marathon runners and walkers. *Journal of Sports Medicine and
       Physical Fitness*. 1964;50:135–41.

[91]   Parker BM, Londeree BR, Cupp GV, Dubiel JP. The noninvasive cardiac evaluation of
       long-distance runners. *Chest*. 1978;73(3): pp. 376–81.

[92]   Douglas P, O'Toole M, Hiller W, Hackney K, Reichek N. Electrocardiographic diagno-
       sis of exercise-induced left ventricular hypertrophy. *American Heart Journal*. 1988;116(3):
       pp. 784–90.

[93]   Sohaib SMA, Payne JR, Shukla R, et al. Electrocardiographic (ECG) criteria for deter-
       mining left ventricular mass in young healthy men; Data from the LARGE Heart study.
       *Journal of Cardiovascular Magnetic Resonance: Official Journal of the Society for Cardiovascular
       Magnetic Resonance*. 2009;11:2.

[94]   Ryan M, Cleland J, French J, et al. The standard electrocardiogram as a screening test for
       hypertrophic cardiomyopathy. *The American Journal of Cardiology*. 1995;76:689–94.

[95]   Bianco M, Bria S, Gianfelici A, et al. Does early repolarization in the athlete have analogies
       with the Brugada syndrome? *European Heart Journal*. 2001;22(6): pp. 504–10.

[96]   Gibbons LW, Cooper KH, Martin RP, Pollock ML. Medical examination and electro-
       cardiographic analysis of elite distance runners. *Annals of the New York Academy of Sciences*.
       1977;301: pp. 283–96.

[97]   Uberoi A, Stein R, Perez MV, et al. Interpretation of the electrocardiogram of young ath-
       letes. *Circulation*. 2011;124(6): pp. 746–57.

[98]   Paetsch I, Reith S, Gassler N, Jahnke C. Isolated arrhythmogenic left ventricular cardio-
       myopathy identified by cardiac magnetic resonance imaging. *European Heart Journal*. 2011:
       DOI: 10.1093/eurheartj/ehr240.

[99]   Marcus FI. Prevalence of T-wave inversion beyond V1 in young normal individuals and
       usefulness for the diagnosis of arrhythmogenic right ventricular cardiomyopathy/dysplasia.
       The American Journal of Cardiology. 2005;95(9): pp. 1070–1.

[100] Papadakis M, Basavarajaiah S, Rawlins J, et al. Prevalence and significance of T-wave inver-
       sions in predominantly Caucasian adolescent athletes. *European Heart Journal*. 2009;30(14):
       pp. 1728–35.

[101] Choo JK, Abernethy WB, Hutter AM. Electrocardiographic observations in professional
       football players. *The American Journal of Cardiology*. 2002;90(2): pp. 198–200.